The JCT Minor Works Building Contracts 2005

Fourth Edition

David Chappell

Blackwell Publishing editorial offices:
Blackwell Publishing Ltd, 9600 Garsington Road, Oxford OX4 2DQ, UK
Tel: +44 (0)1865 776868
Blackwell Publishing Inc., 350 Main Street, Malden, MA 02148-5020, USA
Tel: +1 781 388 8250
Blackwell Publishing Asia Pty Ltd, 550 Swanston Street, Carlton, Victoria 3053, Australia
Tel: +61 (0)3 8359 1011

First edition published 1990 by Legal Studies and Services (Publishing) Ltd
Second edition published 1999 by Blackwell Science Ltd
Third edition published 2004 by Blackwell Publishing Ltd
Fourth edition published 2006

ISBN-13: 978-14051-5271-6
ISBN-10: 1-4051-5271-0

Library of Congress Cataloging-in-Publication Data is available

A catalogue record for this title is available from the British Library

Set in 10/12.5pt Palatino
by TechBooks International, India
Printed and bound in Great Britain
by TJ International, Ltd, Padstow, Cornwall

The publisher's policy is to use permanent paper from mills that operate a sustainable forestry policy, and which has been manufactured from pulp processed using acid-free and elementary chlorine-free practices. Furthermore, the publisher ensures that the text paper and cover board used have met acceptable environmental accreditation standards.

For further information on Blackwell Publishing, visit our website:
www.blackwellpublishing.com/construction

Contents

Preface to the Fourth Edition

The Agreement for Minor Building Works has been the most widely used of the JCT forms of contract. Since 1968, when it was first issued, it has been used consistently not only for simple short contracts of moderate price for which it is intended, but also for much larger projects for which it was often not suited at all. It was a very short form and no doubt its brevity accounted for much of its popularity. Together with other JCT contracts, it was entirely revised in 2005. It was renamed the Minor Works Building Contract and there is an edition which allows for contractor's design (referred to as MW and MWD respectively). There is every reason to believe that its popularity will continue.

The apparent simplicity of this contract was, and still is, deceptive. Things omitted may be as significant as the express terms included. In the preface to the first edition, it was remarked that, like all forms of contract, it must be read against the considerable and growing body of case law. Unfortunately, the case law continues to grow with little equivalent dropping off of cases which are no longer relevant.

This book explains the practical applications of the new version of this contract form from the points of view of the employer, the architect and the contractor. It is not a clause-by-clause analysis. The form has been considered under topic headings where the effect of various clauses can be gathered together. Legal language has been eschewed in favour of simple explanations of legal concepts supported by flowcharts, tables and sample letters. Where possible, I have attempted to provide a clear statement of the legal position in a variety of common circumstances together with references to decided cases so that those interested can read further. Occasionally, where the precise legal position is not clear, a view has been offered. This book is intended to be a practical working tool for all those using MW and MWD.

Over a decade and a half has passed since the first edition was drafted. During that time, much has changed. This edition has been heavily revised to deal with the complete rewriting and re-issue of the MW 98 contract which took place in 2005 effectively producing two contracts, one traditional as before (MW) and the other catering for those instances when the contractor is required to undertake some design work as well as construction (MWD). There is some new terminology in the contracts and the clauses have been substantially reorganised and reworded. Contract particulars and schedules have been added. Account has been taken of relevant new cases and of the April 2004 editions of the RIBA terms of engagement, SFA/99, CE/99 and SW/99.

The first edition was written with the eminent contract authority, the late Professor Vincent Powell-Smith. Such is the rate of change in contracts, case law and legislation, that sadly little if anything now remains of his original contribution.

David Chappell
Wakefield
July 2006

Note: Throughout the text, the contractor has been referred to as 'it' on the basis that it is a corporate body.

CHAPTER ONE
THE PURPOSE AND USE OF MW AND MWD

1.1 *The background*

The JCT Agreement for Minor Works was first published in 1968; it was revised in 1977. The headnote explained that it was intended for minor building works or maintenance work, based on a specification or a specification and drawings, to be carried out for a lump sum. It was inappropriate for use with bills of quantities or a schedule of rates.

Despite its shortcomings, it was widely used for small projects and even for larger ones, its main attraction being its brevity and apparent simplicity. The evidence suggests that architects were becoming increasingly dissatisfied with the length and complexity of the then current main JCT standard form contract (JCT 63) and then – as now – wished to use simple contract conditions wherever possible.

The Minor Works form was extensively revised by a JCT working party in 1979 and a new edition was published in January 1980 and reprinted with corrections in October 1981. It was revised again in April 1985, January 1987, March and October 1988, September 1989, January 1992 and March 1994. Amendment MW9:1995 was issued to take account of the Construction (Design and Management) Regulations 1994 and Amendment MW10:1996 dealt with more insurance changes; Amendment MW11:1998 made significant changes arising out of the Latham Report and the Housing Grants, Construction and Regeneration Act 1996. In effect, MW 80 in its revised form was a completely new set of contract conditions. The headnote to the form as issued in 1980 set out its purpose:

> 'The Form of Agreement and Conditions is designed for use where minor building works are to be carried out for an agreed lump sum and where an Architect or Contract Administrator has been appointed on behalf of the Employer. The Form is not for use for works for which bills of quantities have been prepared, or where the Employer

wishes to nominate sub-contractors or suppliers, or where the duration is such that full labour and materials fluctuations provisions are required; nor for works of a complex nature or which involve complex services or require more than a short period of time for their execution.'

Users found this headnote misleading and it was withdrawn in August 1981 and replaced by Practice Note M2, which is much more indicative of the scope of the contract.

A new form of contract was printed at the end of 1998. It was based on the 1980 edition with JCT Amendments MW1 to MW11, together with some changes and corrections. The new form was still recognisably derived from MW 80 and it was referred to as MW 98. There were a further five amendments to MW 98 since publication. In 2005, the whole suite of JCT contracts was revised and MW 98 became MW with a variant (MWD) incorporating a contractor's designed portion (CDP). For work carried out in Northern Ireland a short adaptation schedule is available.

1.2 *The use of MW and MWD*

The criteria for use of the form are set out inside the front cover of the form. They are set out below with comments:

- Where the work involved is simple in character. The form is relatively short and it is not sufficiently detailed for use where anything complex, whether in the structure of the building itself or in the services, is envisaged. More complex building often raise issues of valuation, extensions of time and financial claims. Even with the terms which the law will imply into this contract, it is not suitable for complex work.
- Where the work is designed by or on behalf of the employer. In the MWD, unsurprisingly, there is an addition dealing with the situation where the contractor is required to design part of the work. MWD is a much needed alternative. Some architects and quantity surveyors believe that contractor's design can be imported by means of a carefully inserted clause in the specification; not so.
- Where the employer is to provide drawings and specification or work schedules to define the quantity and quality of the work. It should be noted that the contractor's obligation is to carry out the work shown collectively in the contract documents. There is no provision

for guaranteed quantities as in some of the forms intended for larger Works.
- Where it is intended that a contract administrator is to administer the contract. The phrase 'contract administrator' is sufficiently wide to include any person so designated by the parties. In theory, this could even be the employer, but there are problems with that approach.

The contract is said not to be suitable if bills of quantities are required, although often the work schedules appear to be bills of quantities by another name. It is not suitable for use if it is desired to have some of the work carried out by named specialists, because there are no clauses to govern the process. Clearly, the use of nominated or named sub-contractors as such would require substantial amendments to the form as printed, although MW envisages that the contractor may sub-let with the architect's consent. It has sometimes been suggested that the control of specialists can be achieved by means of the employer contracting directly with the specialist. This may have other unfortunate consequences. Possible ways of dealing with the situation are explored in Chapter 8.

The contract is not suitable where detailed control procedures are needed, because there are no detailed procedures. Detailed control would be needed for a complex building. MW is not suitable if the contractor is to design part of the Works; MWD should be used. However, MWD cannot be used as a design and build contract. That is a pity, because the industry is sorely in need of a design and build contract for simple work.

The criteria no longer, as in former editions, give advice about the value of contracts for which MW is suitable. The value was never as important as the simplicity of the work and the contract period. No period is suggested either. Very tentatively, an upper limit of about £150,000 and a contract period no longer than six months might be suggested so far as the current form is concerned.

MW should not be used merely on account of its apparent simplicity or because, sensibly, the architect dislikes the complex administrative procedures of the Standard Building Contract (SBC) or the more detailed provisions of the Intermediate Building Contract (IC). The brevity and simplicity of MW is more apparent than real because its operation depends to a large extent on the gaps in it being filled by the general law. Some of the more obvious gaps can be plugged by drafting (or getting an expert in construction contracts to draft) suitable clauses. To take but one example: nowhere in MW is there any provision dealing with contractor's 'direct loss and/or expense' claims as found in SBC and

IC, except for the provisions of clause 3.6.3 which require the valuation of variations to include any direct loss and/or expense incurred by the contractor due to regular progress of the Works being affected by compliance with a variation instruction.

This does not mean that the contractor must allow for the possibility of claims in its tender price or that, in appropriate circumstances, it cannot recover them. It merely means that there is no contractual right to reimbursement and that the architect has no power under the contract to deal with them. As explained in Chapter 10, the contractor can pursue such claims in adjudication, arbitration or by means of legal action. In Chapter 10 there are appropriate suggestions for dealing with this familiar construction industry problem.

1.3 Arrangement and contents of MW and MWD

MW consists of only seven (previously eight) contract conditions, subdivided in what is not always the most logical way. In addition, there are schedules and contract particulars introduced for the first time. The basic form now consists of 32 pages. The form is not for use in Scotland which has its own legal system and applicable rules of law. The actual conditions of contract occupy 14 pages of the document. The whole document is arranged as follows:

ARTICLES OF AGREEMENT
RECITALS
ARTICLES
CONTRACT PARTICULARS
ATTESTATION
CONDITIONS
1 Definitions and interpretation
  Definitions (1.1)
  Agreement etc. to be read as a whole (1.2)
  Headings, references to persons, legislation etc.(1.3)
  Reckoning periods of days (1.4)
  Contracts (Rights of Third Parties) Act 1999 (1.5)
  Giving or service of notices and other documents (1.6)
  Applicable law (1.7)
2 Carrying out the Works
  Contractor's obligations (2.1)
  Materials, goods and workmanship (2.2 in MWD only)
  Commencement and completion (2.2 (2.3 in MWD))
  Architect/Contract Administrator's duties (2.3 (2.4 in MWD))

Correction of inconsistencies (2.4 (2.5 in MWD))
Divergences from Statutory Requirements (2.5 (2.6 in MWD))
Fees or charges legally demandable (2.6 (2.7 in MWD))
Extension of time (2.7 (2.8 in MWD))
Damages for non-completion (2.8 (2.9 in MWD))
Practical completion (2.9 (2.10 in MWD))
Defects (2.10 (2.11 in MWD))
Certificate of making good (2.11 (2.12 in MWD))

3 Control of the Works
Assignment (3.1)
Person-in-charge (3.2)
Sub-letting (3.3)
Architect/Contract Administrator's instructions (3.4)
Non-compliance with instructions (3.5)
Variations (3.6)
Provisional sums (3.7)
Exclusion from the Works (3.8)
CDM Regulations – undertakings to comply (3.9)
Appointment of successors (3.10)

4 Payment
VAT (4.1)
Construction Industry Scheme (CIS) (4.2)
Progress payments and retention (4.3)
Failure to pay amount due (4.4)
Penultimate certificate (4.5)
Notices of amounts to be paid and deductions (4.6)
Contractor's right of suspension (4.7)
Final certificate (4.8)
Failure to pay final amount (4.9)
Fixed price (4.10)
Contribution, levy and tax charges (4.11)

5 Injury, damage and insurance
Liability of contractor – personal injury or death (5.1)
Liability of contractor – injury or damage to property (5.2)
Contractor's insurance of his liability (5.3)
Insurance of the Works by contractor (5.4A)
Insurance of the Works and any existing structures by the employer – fire etc. (5.4B)
Evidence of insurance (5.5)

6 Termination
Meaning of insolvency (6.1)
Notices under section 6 (6.2)
Other rights, reinstatement (6.3)

There is, happily, little supporting documentation required. It consists only of a guidance note, four pages long, bound into the back of the form.

1.4 *Contractual formalities*

Like any other contract, a building contract is made by an offer and acceptance and supported by consideration. The contractor's tender is the offer and a properly worded letter of acceptance from the employer (or from the architect on the employer's behalf) will create a binding contract. Acceptance of the contractor's tender must be unqualified. An acceptance subject to further agreement does not result in a contract.

It is sensible for a formal contract to be executed by the parties, and it is best to use the printed form for this purpose. In the private sector the preparation of the contract documentation is the architect's responsibility, and great care must be taken.

Two copies of the printed form are required, and these must be completed identically, filling in the blanks and making the necessary alterations and deletions. The printed form as completed will then be signed or executed as a deed by the employer and the contractor. This has been simplified by the addition of attestation pages.

The articles of agreement

Page 1 should be completed with the descriptions and addresses of the employer and the contractor. The date will not be inserted until the form is signed or otherwise executed.

The first recital is vital because it is here that a description of 'the Works' must be inserted, and that description is important when considering the question of variations, and for other purposes as well. Only two and a half lines are allowed for this description, so the description must be both precise and succinct. The question of contract documents is discussed in Chapter 3 and the necessary deletions must be made in the second recital.

The second recital is declaratory, and the contract drawings must be signed by or on behalf of the parties using a simple formula such as: 'This is one of the contract drawings referred to in the Agreement made on [date] between [the employer] and [the contractor]'.

The third recital deals with pricing and makes provision for the contractor to price the specification, schedules or to provide a schedule of rates, and it should be appropriately completed. If the architect has chosen to go out to tender on the basis of drawings and the contract conditions only and the contractor also submits a schedule of rates, the architect would be very unwise to let the reference to the schedule of rates stand. It is not required and its existence should not be acknowledged.

It is also necessary to complete Articles 2, 3, 4 and 5.

Article 2 sets out the contract sum and this should be written out in full with the figures in brackets. Of course this sum may be increased or decreased as a result of the operation of the variation clause and if fluctuations are applicable, this sum will be subject to the provisions of schedule 2. That is what is meant by the words 'such other sum as shall become payable'.

Article 3 defines the term 'Architect/the Contract Administrator' and the architect must insert a name and address (or that of the firm) in the blank space. The title 'Contract Administrator' is to cater for the situation where the person nominated is not registered under the Architect's Act 1997. For brevity, this book will refer to that person as 'architect'. MW provides for a designated architect and the contract is in fact inoperable without an architect who is charged with the performance of many important duties under the contract terms, as discussed in Chapter 4. Indeed, so important is the architect's role, that the employer *must* appoint a successor architect if the named architect retires, resigns the appointment or dies. Article 3 requires the employer to do this within 14 days of the original architect's ceasing to act. The successor architect cannot disregard or overrule any certificate or instruction previously given. In some instances, an employer has been known to nominate him- or herself as replacement. Where the original nomination was an architect, such an action is clearly unacceptable. Even where the person was not an architect, it is likely to be implied that a replacement should be of equivalent qualifications to the original, for example building surveyor, engineer, etc., given the importance of the role and the fact that the contractor originally tendered on the basis of a suitably qualified person.

Articles 4 and 5 deal with the effects of the Construction (Design and Management) Regulations 1995. Article 4 assumes that the architect will be the planning supervisor. The user of the contract can insert an alternative name. Article 5 states that the contractor will be the principal contractor. In the event of either planning supervisor or principal contractor ceasing to act, the planning supervisor and/or the principal contractor will be whoever the employer appoints in accordance with regulation 6(5) of the CDM Regulations. All deletions or alterations on the recitals, articles of agreement and, indeed, in the contract conditions must be struck through and initialled by the employer and the contractor.

Article 6 deals with adjudication. Article 7 deals with arbitration. Article 8 deals with legal proceedings. If arbitration is required, Article 8 should be deleted. If legal proceedings are required, Article 7 should be deleted.

Importantly, any such decision should be recorded in the appropriate place in the contract particulars. The user should remember that the default position, if no preference is expressed for either arbitration or legal proceedings, is that disputes will be resolved by legal proceedings. This is a major change in all the JCT contracts from the former position that defaulted to arbitration. Arbitration has always been the method of choice for dispute resolution in the construction industry (as in many others) and there is much to be gained by the appointment of an arbitrator who is experienced in the industry. It seems, however, that lawyers are more comfortable with legal proceedings through the courts which, despite the improvements wrought by the Civil Procedure Rules, may still be thought to offer little to employers and contractors who wish their differences to be resolved by someone with a real knowledge of construction.

Contract particulars

MW 98 had no appendix such as in larger JCT contracts. Details of dates for commencement and completions and the rate of liquidated damages were to be inserted in the text. This was never a good system and it was all too easy, and unfortunately frequent, for items to be missed. In common with other JCT 2005 series contracts, MW and MWD have a section after the articles called 'Contract Particulars' in which all the variables are gathered. Time must be allocated so that the contract particulars can be filled in with great care.

Most of the entries are self-evident, but all must be read carefully before completion. In respect of Article 7, the item must be completed so as to state whether Article 7 and schedule 1 apply (i.e. *not* legal proceedings) or do not apply (i.e. legal proceedings apply instead). Failing to complete this item will result in the system of dispute resolution being legal proceedings.

Another important difference from the former MW 98 is that there is provision for the adjudicator to be named at the commencement of the contract, but failing that or if the named adjudicator cannot act, the adjudicator is to be nominated by any one of the nominating bodies listed. It is the choice of the party wishing to refer a dispute to adjudication. Previously, all the nominating bodies on the list had to be deleted save one and, if no deletions were made, there was a default provision to the Royal Institute of British Architects. The default provisions still apply to the appointor of an arbitrator.

Attestation

A page has been included to allow MW easily to be executed as a deed if desired. It is no longer necessary to seal it. The Companies Act 1989 and the Law of Property (Miscellaneous Provisions) Act 1989 set out the requirements. This is a decision which the employer will have made – on the architect's advice – at pre-tender stage, and there is an important practical difference.

The Limitation Act 1980 specifies a limitation period – the time within which an action may be commenced – of six years where the contract is merely signed by the parties ('a simple contract') or of twelve years where the contract is made as a deed (a specialty contract'). The longer period is beneficial from the employer's point of view and that is why most local and public authorities insist on building contracts being executed as deeds.

Most building contractors operate as limited companies and, by law, they are required to have a company seal which they can use if desired, but it will not alone create a deed. In addition, the document must state on its face that it is a deed. Stamp duty is no longer payable on building contracts executed as deeds unless they are complicated by certain matters such as conveyances or leaseback, which will not normally be the case with MW.

Whichever method is used, it is the architect's duty to check over the contract and make sure that all the formalities are in order. All too often these administrative chores are pushed on one side and then when something goes wrong the legal profession has a field day.

Table 1.1 tabulates the decisions which have to be made on the contract clauses. All necessary decisions as to the applicable clauses and amendments must be made at pre-contract stage, and will have been notified to the contractor accordingly. The contractor is entitled to know the terms on which it is expected to contract. In completing the printed form, the architect must ensure that the final version for execution accords with the terms on which the contractor submitted its tender.

1.5 Problems with the contract documents

Problems can arise with the contract documents which have been prepared for signature or execution as a deed by the architect. The contractor must check them over carefully, checking the printed contract against the information given in the tender documents and noting any discrepancies.

If the contractor discovers mistakes or inconsistencies, it should not execute the documents until the matter is rectified. Figure 1.1 is a suitable pro forma letter for the contractor to send to the architect.

Another problem arises where, as is not uncommon, work starts on site before the contractual formalities are completed. It is bad practice to allow this to happen.

It may be that there is already a binding contract in existence, the parties having agreed on the minimum essential terms and there being an unequivocal acceptance by the employer of the contractor's tender. Alternatively, there may be no contract at all, and in that event and if no contract came into being, the work done would have to be the subject of a *quantum meruit* ('as much as it is worth') claim and the contractor would be entitled to a fair commercial rate: *Lazerbore* v. *Morrison Biggs Wall* (1993). However, calculating the amount due is often quite complicated: *Serck Controls Ltd* v. *Drake & Scull Engineering Ltd* (2000). In general, if a formal contract is executed subsequently, its terms would have retrospective effect: *Tameside Metropolitan Borough Council* v. *Barlows Securities Group Services Ltd* (2001).

It is better to avoid the potential difficulties, and a contractor who is asked to start work before the contractual formalities are completed is well advised to write to the architect appropriately: Figure 1.2.

Table 1.1
Filling in the MWD form (MW variations shown)

Item or clause	Comment
Articles of Agreement	Names and addresses of the parties inserted. Date to be inserted when the last party executes the contract.
First recital	The description of the Works must be sufficient to identify them clearly.
Second recital (MWD)	Insert extent of CDP work or refer to and identify attached signed sheet.
Third recital (Second recital under MW)	The contract drawing numbers must be filled in or an attached list of numbers identified. Delete inappropriate documents.
Fourth recital (Third recital under MW)	Delete inappropriate documents.
Article 2	Contract sum in words and figures inserted.
Article 3	Insert the name and address of the architect.
Article 4	Insert the name of the planning supervisor if all the CDM Regulations apply, otherwise delete.
Article 5	If all the CDM Regulations apply, insert the name of the principal contractor if it is not the main contractor, otherwise delete.
Article 6	This may be deleted if the contract is not a 'construction contract' as referred to in the Housing Grants, Construction and Regeneration Act 1996.
Article 8	Amend reference to 'English' to a different jurisdiction if required.
Contract particulars Fifth recital (Fourth recital under MW)	Delete inappropriate reference to CDM Regulations.
Contract particulars Article 7	Delete to show if arbitration is to apply. If no deletion, legal proceedings will apply.
Contract particulars 2.3 (2.2 under MW)	Insert date of commencement. Do *not* insert 'to be agreed'. Insert date for completion. Do *not* insert 'to be agreed'.

Table 1.1 *Contd*

Item or clause	Comment
Contract particulars 2.9 (2.8 under MW)	Insert amount of liquidated damages and the period, e.g. 'day' or week'. Do not use the words 'per week or part thereof'.
Contract particulars 2.11 (2.10 under MW)	The usual period is three months. If a different period is required, it should be inserted.
Contract particulars 4.3	The 95% figure shown is usual. If desired, a different figure may be inserted.
Contract particulars 4.5	The 97½% figure shown is usual. If desired, a different figure may be inserted.
Contract particulars 4.8.1	Three months is shown. If a longer or shorter period is desired, it should be inserted.
Contract particulars 4.11 and schedule 2	Delete if fluctuations are not to apply.
Contract particulars 4.11 and schedule 2, para. 13	Enter the percentage if an additional sum is to be paid, otherwise enter 'Nil'.
Contract particulars 5.3.2	Insert the amount of insurance cover required.
Contract particulars 5.4A and 5.4B	Delete the inappropriate clause.
Contract particulars 5.4A.1 and 5.4B.1	If the percentage to cover professional fees is not to be 15%, insert another figure.
Contract particulars 7.2	Insert name of adjudicator if desired or write 'not used'. Delete four of RIBA, RICS, CC, NSCC, CIArb.
Contract particulars schedules 1 and 2	Insert the date chosen as the base date having considered schedule 1, paragraph 1 and schedule 2, paragraphs 1.1, 1.2, 1.5, 1.6, 2.1 and 2.2.
Contract particulars schedule 1, para. 2.1	Delete two of RIBA, RICS, CIArb.
Clause 1.7	Amend if not the law of England.

Figure 1.1
Letter from contractor to architect if mistakes in contract documents and no previous acceptance of tender

Dear Sir

PROJECT TITLE

We are in receipt of your letter of the [insert date] with which you enclosed the contract documents for us to sign/execute as a deed [*delete as appropriate*].

There is an error on [describe nature of error and page number of document]. This is not consistent with the tender documents on which our tender is based and we are not prepared to enter into a contract on the basis of the contract documents in their present form.

We therefore return the documents herewith and we look forward to receiving the corrected documents as soon as possible.

Yours faithfully

Figure 1.2
Letter from contractor to architect if contractor asked to commence before contract documents signed

Dear Sir

PROJECT TITLE

Thank you for your letter of the [*insert date*] from which we note that the employer requests us to commence work on site pending completion of the contract documents.

It is our understanding of the situation that we are already in a binding contract with the employer on the basis of our tender of the [*insert date*] and the employer's acceptance of the [*insert date*] on terms incorporated by such tender and letter of acceptance.

If the employer will send us written confirmation of agreement with our understanding of the situation as expressed in this letter, we will be happy to commence as requested.

Yours faithfully

1.6 *Notices, time and the law*

Clauses 1.4 and 1.6 give effect to sections 115 and 116 of the Housing Grants, Construction and Regeneration Act 1996. If the method of service of a notice or any other document (i.e. a certificate or an instruction) is not specified in the contract, it may be served by any effective means. That is by any means which has the desired effect. It can be served to any agreed address. If the parties, for one reason or another, cannot agree the address in each case, the second part of clause 1.6 comes into play and service can be achieved by delivery or by pre-paid post to the intended recipient's last known main business address. If the recipient is a corporate body, the address should be its registered or principal office. There appears to be nothing to prevent second class postage being employed although it would not be in the sender's interest to do so. The Civil Procedure Rules are only binding in legal proceedings, but they provide a useful set of guidelines. Under the CPR:

First class post:	is deemed served the second day after it was posted.
Delivery by hand:	is deemed served the day after it was delivered or left at the permitted address.
Fax:	is deemed served, if transmitted on a business day before 4PM, on that day; or in any other case, on the business day after the date on which it is transmitted.
Other electronic method:	is deemed served on the second day after the day on which it is transmitted.

If a document is served personally after 5PM on a business day or at any time on a Saturday, Sunday or a bank holiday, it will be treated as being served on the next business day ('business day' being any day except Saturday, Sunday or a bank holiday).

Where anything is stated to be done within a number of days after a certain date, the counting of days begins immediately after that date. Christmas Day and Good Friday are excluded, together with any day which is a bank holiday under the Banking and Financial Dealings Act 1971.

Clause 1.7 states that the law applicable to the contract is the law of England. That is the case whatever may be the nationality of any of the parties to the contract or anyone connected to it. Even if the works are carried out in Germany under this contract (unlikely) or in France (even more unlikely) the applicable law will still be the law of England. Obviously, the parties may change the applicable law, for example to the law of Northern Ireland.

In common with other forms of contract, MW excludes the rights of third parties under the Contracts (Rights of Third Parties) Act 1999. This effectively reinstates the position as it was before the Act came into force, i.e. only parties to the contract can have any rights under it (clause 1.5).

CHAPTER TWO
CONTRACT COMPARISONS

2.1 *Introduction*

Advising the employer on which form of contract is best suited to the particular project is one of the architect's important functions. This used to be clear from the schedule of services of the conditions of engagement (CE/95) where under item E06 it stated:

> 'Advise on and recommend form of building contract and explain the Client's obligations thereunder.'

Surprisingly, this duty is absent from the schedule of services in the more recent CE/99 (2004 update). However, it is thought that the architect's duty to advise the client about the most appropriate contract will certainly be implied unless the client obtains advice from another professional about the matter. In view of the architect's involvement in contract administration, this task is one which should certainly be undertaken.

In the classic statement of the architect's duties – for breach of which an architect may be sued – the sixth duty is put in this way:

> 'To consult with and advise the employer as to the form of contract to be used (including whether or not to use bills of quantities) and as to the necessity or otherwise of employing a quantity surveyor . . . '

This sentence – taken from Hudson's *Building Contracts*, 11th edition, page 266, should be engraved on every architect's heart, and advising the employer on which form of contract to use is one of the most neglected of professional functions. It is likely that an architect would be held professionally negligent if the use of an unsuitable form of contract was advised and, as a direct result, the client suffered loss. The suffering of loss is crucially important. There are many instances where the use of the wrong form of contract has not resulted in a loss. A substantial body of case law has built up as the result of using standard contract forms for purposes for which they were not intended and the architect should certainly explain to the client the main provisions of whichever form of contract is recommended.

As regards the JCT 98 contracts, JCT issued a very generalised practice note (Practice Note 5, Series 2), the most useful feature of which was the comparison which it drew between the range of JCT 98 contracts then available. At the time of writing, the practice note has not been updated.

In the private and local authority sectors, the effective choice for work which has been fully designed by the employer's consultants is between one of three Joint Contracts Tribunal forms, the form produced by the Association of Consultant Architects (ACA3) and the more recent Engineering and Construction Contract (NEC).

The ACA form – a second and vastly improved edition of which was published in 1984 and revised from time to time thereafter, most recently in 2003 – is probably not suitable for truly 'minor Works', but there will be many instances where 'minor Works' shade into 'intermediate Works'.

The essential point about MW is that the contract conditions provide only a skeleton; but certainly where simplicity of administration is desired, the Works are uncomplicated, there is no desire for nominated specialists and bills of quantities are not required, MW should be considered. If contractor design is required for part of the Works, MWD is indicated. It was not unknown for a contract to be let on MW 98 where the value was in excess of £3 million. On any view, that must be wrong, not to say negligent, on the part of the architect.

The critical factors in contract choice are:

- The scope and nature of the work
- The presence or absence of bills of quantities
- Lump sum or approximate price.

In the JCT family – if it is decided that bills of quantities are inappropriate – the effective choice is between the 'without quantities' SBC/XQ, IC and MW. Key features of MW are:

- *Client control*
  The employer must appoint an architect or contract administrator to administer the contract. There is no longer any provision for the appointment of a quantity surveyor and, in any event, no express quantity surveyor duties.

- *Design responsibility*
  There is provision to allow the contractor to carry out and be responsible for design in MWD. This is the equivalent to the contractor's designed portion supplement under the standard SBC form.
- *Possession and completion*
  MW contains a simple provision (Clause 2.2) stating that the Works 'may be commenced' on a specified date. This particular wording is of no real significance in light of the common law position. There is a date for completion. If the contractor defaults in completion, there is a traditional liquidated damages clause. There is no provision for a non-completion certificate.
- *Sectional completion and partial possession*
  There are no provisions for sectional completion or partial possession in MW.
- *Extension of time*
  In contrast to SBC and IC, MW has very brief provisions for extensions of time. The contractor must notify the architect of his need for an extension; and the architect must grant a reasonable extension of the contract time for reasons beyond the control of the contractor. There is no review period.
- *Reimbursement for loss and/or expense*
  Apart from the provision in clause 3.6.3 requiring the architect to include 'direct loss and/or expense' when valuing a variation instruction or as a result of the employer's compliance or non-compliance with clause 3.9, there are no provisions for reimbursing the contractor for 'direct loss and/or expense' – but the contractor can recover under the general law.
- *Selection of sub-contractors*
  Provided that the architect gives consent, the contractor can choose domestic sub-contractors. There are no provisions for naming or nomination of sub-contractors. Some might consider that a plus point.
- *Variations*
  Variations are valued on a 'fair and reasonable basis' – using any priced documents.
- *Testing and opening up*
  There are no express provisions under MW, but it is considered that the power to order testing or opening up will be implied (see section 4.3).
- *Fluctuations*
  Fluctuations for contributions, levy and tax fluctuations only (clause 4.11).

Table 2.1
MW clauses compared with those of SBC and IC

MW clause	Description	SBC clause	IC clause
1	**Definitions and interpretations**	**1**	**1**
1.1	Definitions	1.1	1.1
1.2	Agreement to be read as a whole	1.3	1.3
1.3	Headings, references to persons, legislation, etc.	1.4	1.4
1.4	Reckoning periods of days	1.5	1.5
1.5	Contracts (Rights of Third Parties) Act 1999	1.6	1.6
1.6	Giving or service of notices and other documents	1.7	1.7
1.7	Applicable law	1.12	1.12
2	**Carrying out the Works**	**2**	**2**
2.1	Contractor's obligations	2.1	2.1
2.2	Commencement and completion	2.4	2.4
2.3	Architect's duties	2.9, 2.10, 2.12	2.9, 2.10, 2.11
2.4	Correction of inconsistencies	2.14	2.13
2.5	Divergences from statutory requirements	2.17	2.15
2.6	Fees or charges legally demandable	2.21	2.3
2.7	Extension of time	2.27, 2.28	2.19, 2.20
2.8	Damages for non-completion	2.32	2.23, 2.24
2.9	Practical completion	2.30	2.21
2.10	Defects	2.38	2.30
2.11	Certificate of making good	2.39	2.31
3	**Control of the Works**	**3**	**3**
3.1	Assignment	7.1	7.1
3.2	Person-in-charge	3.2	3.2
3.3	Subletting	3.7, 3.9	3.5, 3.6

Table 2.1 *Contd*

MW clause	Description	SBC clause	IC clause
3.4	Architect's instruction	3.10	3.8
3.5	Non-compliance with instructions	3.11	3.9
3.6	Variations	3.14, 5.2	3.11, 5.2
3.7	Provisional sums	3.16	3.13
3.8	Exclusion from the Works	3.21	3.17
3.9	CDM Regulations – undertakings to comply	3.25	3.18
3.10	Appointment of successors	3.26	3.19
4	**Payment**	**4**	**4**
4.1	VAT	4.6	4.3
4.2	Construction Industry Scheme (CIS)	4.7	4.4
4.3	Progress payments and retention	4.9, 4.10, 4.11, 4.12	4.6, 4.7
4.4	Failure to pay amount due	4.18.1	4.10
4.5	Penultimate certificate	–	4.9
4.6	Notices of amounts to be paid and deductions	4.13	4.8
4.7	Contractor's right of suspension	4.14	4.11
4.8	Final certificate	4.15	4.14
4.9	Failure to pay the final amount	4.15	4.14
4.10	Fixed price	4.21	4.15
4.11	Contribution levy and tax changes	4.21	4.15
5	**Injury, damage and insurance**	**6**	**6**
5.1	Liability of contractor – personal injury or death	6.1	6.1
5.2	Liability of contractor – injury or damage to property	6.2, 6.3	6.2, 6.3
5.3	Contractor's insurance of his liability	6.4	6.4
5.4A	Insurance of the Works by contractor	6.7	6.7

Table 2.1 *Contd*

MW clause	Description	SBC clause	IC clause
5.4B	Insurance of the Works and existing structures by employer – fire etc.	6.7	6.7
5.5	Evidence of insurance	6.4, 6.7	6.4, 6.7
6	**Termination**	**8**	**8**
6.1	Meaning of insolvency	8.1	8.1
6.2	Notices under section 6	8.2	8.2
6.3	Other rights, reinstatement	8.3	8.3
6.4	Default by contractor	8.4	8.4
6.5	Insolvency of contractor	8.5	8.5
6.6	Corruption	8.6	8.6
6.7	Consequences of termination under clauses 6.4 to 6.6	8.7	8.7
6.8	Default by employer	8.9	8.9
6.9	Insolvency of employer	8.10	8.10
6.10	Termination by either party	8.11	8.11
6.11	Consequences of termination under clause 6.8 to 6.10	8.12	8.12
7	**Settlement of disputes**	**9**	**9**
7.1	Mediation	9.1	9.1
7.2	Adjudication	9.2	9.2
7.3	Arbitration	9.3–9.8	9.3–9.8

- *Payment and retention*
  Progress payments to be made by the employer at four-weekly intervals. The retention is not expressly stated to be trust money.

There are, of course, other significant differences between SBC, IC and MW. The architect cannot assume the same powers as are conferred under the more detailed forms – that is not the case – but this should not deter recommendation of MW in those many cases in which it is the most suitable standard form contract.

2.2 *JCT contracts compared*

Once the decision to use the JCT form of contract for employer-designed works, on a lump sum basis and without bills of quantities, is made, the choice is narrowed to the three best-known and mostly widely used of the JCT forms.

Table 2.1 enables readers to see – it is to be hoped at a glance – the position under MW in comparison with SBC (without quantities) and IC.

CHAPTER THREE
CONTRACT DOCUMENTS AND INSURANCE

3.1 Contract documents

3.1.1 Types and uses

Within its limitations of use (see Chapter 1), MW can be used with a wide variety of supporting documents. Taken together, they are termed the contract documents. In principle, they may consist of any documents agreed between the parties to give legal effect to their intentions.

A number of options are set out in the first recital and they may be conveniently considered as follows:

- The contract drawings
- The contract drawings and the specifications priced by the contractor
- The contract drawings and work schedules priced by the contractor
- The contract drawings, the specification and the schedules, one of which is priced by the contractor.

One of these options, together with the Agreement and Conditions annexed to the Recitals, form the contract documents. They must *all* be signed by or on behalf of the parties.

Before embarking on a project, the options must be studied carefully to arrive at the combination which is most suitable for the work. Note that the third recital provides that the contractor must price either the specification or the work schedules or provide its own schedule of rates. Oddly, it appears that the contractor's own schedule of rates does not become one of the contract documents. (The implications of this are discussed in Chapter 11.)

The contract drawings

On very small Works, for which of course MW is very well suited, it may be quite acceptable to go to tender merely on the basis of MW and drawings. This system can be very satisfactory, provided that all the information required by the contractor for pricing purposes is included on the drawings. The architect is not precluded from issuing further information by way of clarification in accordance with clause 2.3 as necessary; indeed it is the architect's duty to do so, but the contractor cannot expect the architect to provide every detail no matter how minute. The contractor is expected to use its own practical experience in constructing the works: *Bowmer and Kirkland Ltd* v. *Wilson Bowden Properties Ltd* (1996).

Since there is no specification or schedules, the contractor will be expected to provide its own schedule of rates. If this is not intended, the third recital must be deleted in its entirety. The significance of the priced document is that it is to be used to value variations, if relevant, under clause 3.6. Reference to the contractor's own schedule of rates is included in this clause even though it is not one of the contract documents. Some commentators advise that the contractor's own schedule of rates should always be excluded, but it is difficult to see the reason for so doing.

The contract drawings and the specification priced by the contractor

In practice, this is a very common way of dealing with small projects. If the specification is to be priced, great care must be exercised in preparing it. This system involves the contractor in taking off its own quantities with reference to both drawings and specification and the cost of so doing is likely to be reflected in the tender figure. The organisational expertise which is incorporated into the specification will determine how useful the priced document will be in the valuation of variations.

The contract drawings and work schedules priced by the contractor

This is the variant of the last system. Since the Works must be specified somewhere, it is likely that the specification element will be incorporated into the schedules. Alternatively, depending on the type of work, it may be feasible to put all the specification notes on the drawings. The schedules will normally be quantified, making this system much easier to price from the contractor's point of view. Indeed, often the schedules are really bills of quantities under another name. The contractor needs to take care, however, that its price is inclusive of everything required to carry out the Works. This is true even if some items are missed off the schedules but shown on the contract drawings. The point can be a difficult one. It is discussed in section 3.1.2: clause 2.4 (correction of inconsistencies).

The contract drawings, the specification and the work schedules, one of which is priced by the contractor

In practice, this combination would be used on larger Works when the work schedules would take the form of bills of quantities. The contractor would then normally price the schedules rather than the specification. Once again, the contractor must take care that it prices for everything the contract requires it to do since even if the work schedules are in the form of bills of quantities, the employer does not warrant their accuracy, neither does the contract provide that the quantities, if given, take precedence over drawings or specification. Thus, if five doors are listed in the schedules, but it is clear the drawings show seven doors, the contractor must price for seven doors and will be taken to have done so.

In principle, a work schedule is always to be preferred over a schedule of rates. The former is capable (or should be capable) of being priced out and added together to arrive at the tender figure. A difficulty may arise because it requires a broad measure of agreement on the method of carrying out the Works. The contractor will have difficulty in pricing a schedule of work if it considers that a totally new approach will show greater efficiency. Some two-stage method of tendering will probably yield best results in such cases when the contractor can be expected to input its suggestions before the schedules are drawn up. Whether two-stage tendering is justified on small projects is another matter.

On the other hand, the figures in a schedule of rates cannot be added together to give the tender figure, and the contractor's own schedule cannot be accepted unless it has justified the calculation of the overall sum from the basis of the schedule. To do otherwise would reduce the valuation of variations to a farce. A schedule of rates is most useful where the total content of the work is not precisely known at the outset. MW can be used in this way, with a little adjustment, but it is better to consider some other forms such as SBC With Approximate Quantities or the Prime Cost Contract (PCC).

The numbers of the contract drawings must be inserted in the space provided in the second recital. On large contracts, when bills of quantities are used, it is usual to designate as contract drawings only those small-scale drawings which show the general scope and nature of the work. Under this contract, however, the situation is very different. The total number of drawings is likely to be relatively small and the contractor will need all of them in order to prepare its tender. Since the contractor's basic obligation (clause 2.1.1) is to 'carry out and complete the Works in a proper and workmanlike manner and in compliance with the Contract Documents' the architect must be sure that the contract documents taken together do cover the whole of the Works. Therefore, the contract drawings must be:

- The drawings from which the contractor obtained information to submit its tender.
- Sufficiently detailed so that, when taken together with the specification and/or work schedules, they include all workmanship and materials required for the project.

The further information which the architect must provide under clause 2.3 may consist of drawings, details and schedules. Provided that they are merely clarifying existing information, there is no financial implication. If, however, they show different or additional or less work or materials than those shown in the contract documents, the contractor will be entitled to a variation on the contract sum. None of the 'further information' constitutes a contract document.

Every contract document must be signed and dated by both parties to avoid any later dispute regarding what is or what is not a contract document. In practice, this means signing every drawing and the cover of the specification and/or schedules. It is suggested that each document is endorsed: 'This is one of the contract documents referred to in the Agreement dated...' or that some other form of words to the same effect be used.

3.1.2 Importance and priority

The contract documents provide the only legal evidence of what the parties intended to be the contract between them. They are, therefore, of vital importance. In the case of dispute, the adjudicator, the arbitrator or the court will look at the documents in order to discover what was agreed.

Clause 2.1.1 lays an obligation upon the contractor to carry out the Works in accordance with the contract documents. The question often arises: what is the position if the documents are in conflict? Clause 1.2 states that the printed form must be read as a whole; nothing contained in the contract drawings or the specification or the schedules will override or modify the printed conditions. If, therefore, the specification were to contain a clause purporting to remove the contractor's entitlement to extension of time due to exceptionally adverse weather, it would be ineffective because of this provision, which has the effect of reversing the normal legal rule of interpretation that type prevails over print. The effectiveness of a clause worded in this way has been upheld in the courts on many occasions: e.g. in *English Industrial Estates Corporation Ltd* v. *George Wimpey & Co Ltd* (1972).

Although the contract effectively sorts out priorities as between the printed form and the other contract documents, it gives no further guidance as far as priorities among the other contract documents are concerned. Clause 2.4 simply states that any inconsistency in or between the contract drawings and the contract specification and the work schedules must be corrected and if such correction results in addition, omission or other change, it must be treated as a variation under clause 3.6.

The particular circumstances of each case will determine exactly how the inconsistency is to be treated. Two main types of inconsistency are common:

(1) Where workmanship or materials are covered in one of the contract documents, but omitted from the other
(2) Where workmanship or materials shown in one of the contract documents are in conflict with what is shown in the other.

This consideration will be confined to a contract based on drawings and specification which the contractor has priced. If a schedule is also included, the principle is the same but the facts may be more complex. The first thing to establish is what the contractor has legally contracted to do. Article 1 puts the situation very clearly:

> 'The Contractor shall carry out and complete the Works in accordance with the Contract Documents.'

Clause 2.1.1 also refers to the 'Contract Documents' and that clearly means the documents noted in the first recital, that is, in this case, the contract drawings, the contract specification and the conditions. If, therefore, the inconsistency is, for example, the fact that a handrail is shown without bracket on the drawings, but brackets are specified in the specification, or vice versa, it will be deemed that the contractor has allowed for the brackets in its price. This is probably the case even if the brackets are not in the specification, which the contractor has priced, but are shown on the drawing. In this example, even if the brackets are not shown or mentioned on either document, it is likely that the contractor must supply brackets (presumably the cheapest it can find to do the job) at no additional cost: *Williams* v. *Fitzmaurice* (1858). Much, however, will depend on any general terms which have been included in the specification. An expression such as 'The whole of the materials whether specifically mentioned or otherwise necessary to complete the Works must be provided by the contractor' would tend to place the responsibility for omissions squarely on the contractor provided that they could be considered 'necessary' to complete the Works. In this case, brackets are obviously necessary to support the handrail.

If the documents are in conflict, the position is rather complicated. Since neither drawings nor specification have priority, it is not clear as to which of them the contractor has had reference in formulating the price. It is tempting to consider that the key document is the specification, since the contractor has priced it. Article 1 and clause 2.1.1 clearly indicate that this is a wrong view of the situation. In pricing the specification, the contractor must have regard to the totality of the documents. It is thought that, if the documents are in conflict, it is for the architect to instruct the contractor as to which document is to be followed in the particular instance, and the contractor is not to be allowed any addition to the contract sum nor is there to be any omission, the contractor being deemed to have included for whichever option the architect chooses.

If, however, the architect solves the problem by omitting the work or changing it to something other than is contained in either of the contract documents, it falls to be valued in accordance with clause 3.6 in the usual way. Although this view may be thought to impose undue hardship on the contractor in certain circumstances, particularly as the inconsistency is due to the architect's oversight, it is the only interpretation that appears to take account of the contract as a whole. The contractor bears a great responsibility to examine the documents thoroughly when pricing. This analysis has an odd result. Although clause 2.4 refers to treatment of the correction as a variation under clause 3.6, it is difficult to envisage any situation in which the contractor would be entitled to have such a variation valued at anything other than a nil amount. If the contract is on the basis of drawings and priced schedules which are in fact fully developed bills of quantities, the situation remains the same. The employer does not warrant the accuracy of the bills in this instance unless a clause is included to that effect. This, of course, is in complete contrast to SBC and points out one of the great dangers to the contractor in using this form for Works of a greater value than that for which it is intended.

3.1.3 Custody and copies

There is no specific provision in the contract regarding the custody of the contract documents and the issue of copies to the contractor. It is probably sensible for the architect to keep the original, but if the employer wishes to have it, the architect should be sure to have an exact copy. Although there is no express requirement in the contract, it is good practice for the architect to make a copy for the contractor and certify that it is a true copy. This can be simply done by binding all the documents together and inscribing the certificate on each document including the drawings. It is sufficient to state 'I certify that this is a true copy of the contract document dated . . . '.

It must be implied in the contract that the architect provides the contractor with two copies of the contract drawings and specification and/or schedules, otherwise the contractor would be unable to carry out the works. It is usual to provide such copies free of charge.

Any further drawings, details or schedules which the architect is to provide under clause 2.3 are not contract documents. They are intended merely to amplify or clarify the information in the contract documents. The architect is obliged only to supply such drawings as are 'necessary for the proper carrying out of the Works'. It is an implied term of the contract that such additional information will be issued at the correct time: *R. M. Douglas Construction Ltd* v. *CED Building Services* (1985) and that the information will be correct: *London Borough of Merton* v. *Stanley Hugh Leach Ltd* (1985). Failure to do so is a breach on the architect's part for which the employer may be responsible: *Penwith District Council* v. *V. P. Developments Ltd* (1999). The contractor would have a legitimate claim for an extension of time without the necessity for any prior application for the information. If the contractor suffers disruption, it may also have a claim for damages for breach of contract which it could pursue at common law.

3.1.4 Limits to use

The contract contains no express terms to safeguard the architect's interests in the drawings and specification and there is no express prohibition on the employer from using the contractor's rates and prices for purposes other than this contract.

The general law, however, covers the position. Under the Copyright, Designs and Patents Act 1988, the architect retains the copyright in drawings and specification (unless expressly relinquished) and neither the contractor nor the employer may make use of them except for the purpose of the project. Strictly, the architect may ask the contractor to return all copies after the issue of the final certificate, but in practice it is seldom worth the trouble to receive a collection of torn, stained and unreadable pieces of paper.

The confidentiality of the contractor's rates and prices is safeguarded by the general rule that a party has a duty not to divulge confidential information to third parties. This is especially true when two parties are bound together in a contractual relationship and to divulge the information would clearly cause harm. The contractor's prices are a measure of its ability to tender competitively and secure work. To divulge its rates to a competitor is a serious matter. In practice, it is not easy for the contractor to ensure, for example, that the quantity surveyor does not make use of its prices to assist in estimating the cost of other contracts, but that is probably of little consequence.

It may be thought prudent to include a clause in the specification, similar to clause 2.8.3 of IC, to cover the limitations on the use of documents. It does no harm to remind the parties of their obligations in this respect.

3.2 *Insurance*

3.2.1 Injury to or death of persons

Under clause 5.1 the contractor assumes liability for and indemnifies the employer against any expense, liability, loss, claim or proceedings arising out of the carrying out of the Works in respect of personal injury or death of any person, except to the extent due to act or neglect of the employer or any person for whom the employer is responsible. The persons for whom the employer is responsible will include anyone employed by the employer and paid direct such as a directly employed contractor, the clerk of works (if any) and the architect.

In practical terms, the employer will be responsible for the injury or death of any person only insofar as the injury or death was caused by the employer or the employer's agent's act or neglect. The contractor retains liability even if the employer is to some degree responsible. In effect, in such circumstances, the employer makes a contribution to reflect the extent of his or her negligence. The employer will then join the contractor as a third party in any action and claim an indemnity under this clause. Under clause 5.3, the contractor must take out and maintain and cause any subcontractor to take out and maintain insurances which must comply with all relevant legislation in connection with claims for personal injury or death of any person under a contract of service or apprenticeship with the contractor and which arises out of and in the course of the person's employment. There is a space in the contract particulars for the insertion of a suitable sum. It is suggested that the architect advises the employer to get the opinion of an insurance broker as to the amount which should be included. Clause 5.5 gives the employer the right to require evidence that the insurance has been taken out, but there is no provision for the employer to take out the necessary insurance and deduct the amount from the contract sum if the contractor defaults. Nonetheless, failure on the part of the contractor to insure is a breach of contract for which the employer could recover damages at common law, provided of course the damage suffered could be properly evidenced. In practice, it is certain that the employer would set off such sums against money payable to the contractor.

It usually falls to the architect to ask the contractor for evidence of insurance. The architect must not fall into the trap of attempting to interpret the wording of the contractor's policy; neither is it sufficient simply to pass the policy to the employer (often referred to as acting like a postbox). Either of these courses of action may result in the architect becoming liable for the giving of bad advice to the employer: *Pozzolanic Lytag* v. *Bryan Hobson Associates* (1999). The architect should pass the insurance documents to the employer with a note to the effect that the architect is not an insurance expert and suggests that the employer seeks the advice of a broker on the adequacy of the policy. The architect could seek the advice, but that is not to be recommended. It is better if the architect keeps such matters at arm's length.

The requirement that the contractor insures is stated to be without prejudice to its liability to indemnify the employer. In the case of an insurance company failing to pay in the event of an accident, the contractor is bound to find the money itself. The requirements of clause 5.3 should be satisfied if the contractor and subcontractors already have general insurance cover in appropriate terms in an adequate amount.

It is always wise for the employer to retain the services of an insurance broker to give advice about the insurance provisions of the contract. The broker inspects all the relevant documents and certifies to the architect that they comply with the contract requirements. It is essential that the insurance is in operation from the time the contractor takes possession of the site.

3.2.2 Damage to property

Under clause 5.2, the contractor assumes liability for, and indemnifies the employer against any expense, liability, loss, claim or proceedings arising out of the carrying out of the Works in respect of injury or damage to any kind of property other than the Works, any unfixed materials and goods on the Works and any property insured under clause 5.4B to the extent that it is due to the negligence, breach of statutory duty, omission or default of the contractor or any person engaged by the contractor upon the Works. The contractor's liability under this clause is limited compared with its liability under clause 5.1. The contractor or subcontractor must be at fault for the indemnity to operate. The contractor need not be totally at fault for indemnity purposes. If it is partly at fault, the employer has a partial indemnity. The contractor must insure and cause any subcontractor to insure against his liabilities under clause 5.3.

The very instructive case of *National Trust for Places of Historic Interest or Natural Beauty* v. *Haden Young* (1994) provided a clear and sensible explanation of the way the predecessor of this clause should work in connection with clause 5.4B. The case concerned an earlier version of the contract, and it is now clear that the contractor gives no indemnity under clause 5.2, nor does it insure, in respect of the Works, or existing buildings to which the work is being carried out or the contents of such existing buildings. If damage is caused to the Works through the contractor's fault, the contractor may be liable, because its obligation is to carry out and complete the Works in accordance with the contract documents.

In general the position appears to be that if the existing building and contents are the subject of damage by perils listed in clause 5.4B, they should be covered by the insurance under clause 5.4B even if wholly due to the contractor's negligence. Damage caused by matters other than the perils in clause 5.4B must be rectified at the expense of the contractor if due to its negligence, otherwise it is at the risk of the employer. It has been held that a fire caused by the contractor's negligence is not covered by the employer's insurance and must be dealt with under the contractor's own insurance: *London Borough of Barking & Dagenham* v. *Stamford Asphalt Co* (1997). However, that was under a differently worded MW 80.

The contract provides for the employer to stipulate the amount of cover required. The general comments with regard to inspection of documents in section 3.2.1 are also applicable to this clause.

There is no provision in MW for insurance against claims arising due to damage caused by the carrying out of the Works when there is no negligence or default by any party. Most standard forms of contract contain such provision which can be invoked if required, to cover specific risks, for example the carrying out of underpinning works to adjoining property. It is not always necessary to take out such insurance and the provision is probably omitted from the contract in view of the minor nature of the works envisaged. The amount of the contract sum, however, is no indication of the possible risk to neighbouring premises and the architect would be prudent to assess each project on its own merits.

There is no reason why the insurance should not, and every reason why it should, be taken out by the employer with the advice of a broker if circumstances appear to warrant it.

3.2.3 Insurance of the Works against fire etc.

Clause 5.4 deals with the insurance of the Works against a collection of specific risks (referred to as 'Specified Perils'). They are as follows:

> 'fire, lightning, explosion, storm, flood, escape of water from any water tank, apparatus or pipe, earthquake, aircraft and other aerial devices or articles dropped therefrom, riot and civil commotion but excluding Excepted Risks.'

The clause is divided into two parts, only one of which is to apply. Each part is applicable to a particular situation as follows:

(1) 5.4A: A new building where the contractor is required to insure
(2) 5.4B: Alterations or extensions to existing structures where the employer is required to insure.

The contract particulars must be completed to indicate which part is to apply.

3.2.4 A new building where the contractor is required to insure

The provision is not complex and provides for the contractor to insure in the joint names of the employer and itself against loss or damage caused by any of the specified perils. The insurance must cover the full reinstatement value and include:

- All executed Works
- All unfixed materials and goods intended for, delivered to, placed on or adjacent to the Works and intended for incorporation

and exclude:

- Temporary buildings, plant, tools and equipment owned or hired by the contractor or his subcontractors. (This is clear although the contract does not expressly refer to it.)

A percentage is to be added to take account of all professional fees.

Great care must be taken in determining the level of insurance cover which must be inserted in the contract particulars. Provision must be made for the possibility that the Works are virtually complete at the time of the damage and allowance made for clearing away before rebuilding. Although it is the contractor's responsibility, the architect should satisfy him- or herself that the level of cover is adequate by drafting a letter for the employer to send to the contractor under clause 5.5 and if not satisfied, the architect must notify the employer and the contractor immediately (Figures 3.1 and 3.2). Despite the absence of express provision, the employer must insure himself if the contractor defaults or fails to insure adequately. It should be noted that the contractor's 'All Risks' policy will not necessarily insure against all the employer's perceived losses, such as loss of revenue. Depending on circumstances, the contractor may not be liable for such losses in any event: *Horbury Building Systems Ltd* v. *Hampden Insurance NV* (2004).

The insurance must be kept in force until the date of issue of the certificate of practical completion. This is the case even if practical completion is delayed beyond the date for completion in the contract. If the contractor already maintains an 'All Risks' policy which provides the same type and degree of cover required under clause 5.4A, it will serve to discharge the contractor's obligations provided it recognises the employer as joint insured and the contractor must produce documentary evidence at the employer's request. The architect must not simply pass the documents to the employer without comment. As noted earlier, the architect must do one of three things:

- Give advice to the employer, something which the architect is unlikely to be qualified to do; or
- Obtain expert advice on the policy and pass this on to the employer; or
- Advise the employer to obtain expert advice.
  (*Pozzolanic Lytag Ltd* v. *Bryan Hobson Associates* (1999))

If damage occurs, as soon as the insurers have carried out any inspection they may require, the contractor must 'with due diligence' restore or replace work or materials or goods damaged, dispose of debris before proceeding to carry out and complete the Works as before. The contractor is entitled to be paid all the money received from insurance less only the percentage to cover professional fees. It is usual for the money to be released in instalments on certificates, but the contract is silent on the point and it is open to the parties to make any

Figure 3.1
Letter from architect to contractor if contractor fails to maintain an adequate level of insurance under clause 5.4A

Dear Sir

PROJECT TITLE

I refer to my telephone conversation with your [*insert name*] this morning and confirm that your response to the employer's letter of the [*insert date*] shows that you are not maintaining an adequate level of insurance as required by clause 5.4A of the Conditions of Contract.

In view of the importance of the insurance and without prejudice to your liabilities under clause 5.4A the employer is arranging to take out the appropriate insurance on your behalf. Any sum or sums payable by the employer in respect of premiums will be deducted from any monies due or to become due to you or will be recovered from you as a debt.

Yours faithfully

Copy: Employer
Quantity surveyor (if appointed)

Figure 3.2
Letter from architect to employer if contractor fails to maintain an adequate level of insurance under clause 5.4A

Dear Sir

PROJECT TITLE

I am not satisfied that the contractor is maintaining an adequate level of insurance as required by clause 5.4A of the Conditions of Contract.

In view of the importance of the insurance, it may be wise to take out the necessary insurance on the contractor's behalf without delay. To this end, I have already advised your broker of the situation and you should contact him immediately. Although the terms of the contract make no express provision for you to act on the contractor's default, it is my opinion that you are entitled to set off the cost of the premium from your next payment to the contractor in this instance.

A copy of my letter to the contractor, dated [*insert date*], is enclosed for your information.

Yours faithfully

other mutually agreeable arrangement. The contractor is not entitled to any other money in respect of the reinstatement and if there is any element of underinsurance or excess, it must bear the difference itself.

3.2.5 Alterations or extensions to existing structures

In the case of an existing building, the employer must insure the Works in joint names, the existing structures and contents and all unfixed materials as before. The contractor's temporary buildings etc. are again excluded although not expressly. There is a proviso that the employer need only insure such contents as are owned by him or her, or for which the employer is responsible. The reason for this is obscure and it takes little imagination to foresee problems arising. There is now a requirement that the employer insure for the value of professional fees. They are the employer's responsibility and it would be wise to do so. It would be advisable for the architect to suggest the inclusion of a suitable percentage.

The relationship between the predecessors of clauses 5.2 and 5.4B was considered in *National Trust for Places of Historic Interest* v. *Haden Young* (1994). The crucial question was whether the predecessors of clauses 5.2 and 5.4B provided a scheme whereby loss to be insured by the employer fell upon the employer under a different wording; the trial judge emphatically said 'No' and the Court of Appeal agreed with him.

Essentially, clause 5.2 makes the contractor liable for damage to property, except the Works which are the subject of the building contract and property insured under clause 5.4B, to the extent that the damage is due to the contractor's or any subcontractor's negligence or default. The contractor will be liable to a partial extent if it is partially at fault and damage is caused to surrounding buildings, passing vehicles or an existing building to which the Works are being carried out. The Works are expressly excepted, but if damage is caused to the Works themselves through the contractor's fault, the contractor may be liable, because its obligation is to carry out and complete the Works in accordance with the contract documents. The contractor is obliged to carry appropriate insurance.

If the Works are alterations or extensions to an existing structure, the employer is obliged to insure the Works and the existing structure under the provisions of clause 5.4B. Effectively, the employer and contractor agree that if there is damage caused by specified perils, this clause would deal with the situation even if due to the contractor's negligence.

The contractor is entitled to require the employer to produce such evidence as it may need to satisfy itself that the insurance is in force. Local authorities are expressly excused. There is no provision for the contractor to take out insurance itself if the employer defaults, nor to have the amounts of any premiums it pays added to the contract sum. Clearly, however, failure to insure on the part of the employer is a breach of contract, but not one which entitles the contractor to terminate its employment in accordance with clause 6.8 (see section 12.3.2) nor, it is thought, to treat the contract as repudiated at common law. The architect must not allow the situation to arise and must remind the employer at the appropriate time if insurance is to be the employer's responsibility. Although this insurance is clearly not the responsibility of the architect, the architect probably has a duty to advise the employer that such insurance must be taken out.

If any loss or damage does occur, the architect is to issue instructions regarding the reinstatement and making good in accordance with clause 3.4. If there is any element of under-insurance, this time it is the employer who must make good the difference. The contractor is entitled to be paid for the work it carried out in the normal way, i.e. following valuation under clause 3.6.

It must be noted that whether the employer or the contractor is responsible and has to insure, *all* damage to the Works by fire etc. is removed from the contractor's indemnity under clause 5.2.

The employer's obligation to insure in joint names under clause 5.4B ceases on practical completion and the contractor if open to claims for damages from the employer: *TFW Printers Ltd* v. *Interserve Project Services Ltd* (2006).

Unlike the provisions of similar clauses in SBC and IC, there is no provision in this contract for either party to terminate the contractor's employment if it is just and equitable to do so after loss or damage. This is presumably a conscious decision on the part of the JCT, and the position is left to be governed by the general law. The total destruction of the Works by fire etc. might bring the contract to an end at common law, but in the majority of cases the parties should be able to come to some mutually acceptable arrangement depending on all the circumstances.

A more serious omission is that there is no optional clause to cover the situation where there is damage caused to adjacent property which is not due to any negligence on the part of the contractor or the employer. If the architect believes that there is danger of such damage, it is wise to insert a suitable clause.

3.3 Summary

Contract documents

- They may consist of any documents agreed between the parties
- Taken together, they must cover the whole of the work
- The information provided under clause 2.3 is additional to and not part of the contract documents
- They are the only legal evidence of the contract
- The printed conditions override anything contained in the drawings or specifications
- The correction of inconsistencies may not result in a variation
- The architect has an implied duty to supply the contractor with sufficient copies of the contract drawings and specification and/or schedules for the contractor to carry out and complete the work
- Further correct information necessary must be issued at the correct time or the contractor may claim an extension of time and damages at common law
- No drawing or specification must be used for any purpose other than the contract and the contractor's prices must not be divulged to third parties.

Insurance

- Employer's indemnity covers personal injury and death and damage to property – subject to certain exceptions
- Contractor must insure to cover the indemnities
- The employer has no right to insure if the contractor defaults
- There is no provision for insurance to cover damage to property not the fault of either party
- New work must be insured by the contractor against the specified risks
- Alterations and extensions and existing work must be insured by the employer against the specified risks.

CHAPTER FOUR
ARCHITECT

4.1 *Authority and duties*

The contract contains express provisions regarding the extent of the architect's authority and the duties imposed upon him. These provisions are discussed in detail below (section 4.2). It would be quite wrong to think, however, that the provisions specifically set out in the contract are the end of the matter. The contract is very brief and, even in the case of a much more comprehensive contract form such as SBC, the architect has obligations which are sensible but not always immediately apparent.

From the contractor's point of view, the architect's authority is indeed defined by the contract. Thus an architect who attempts to exceed this authority may, quite rightly, be ignored by the contractor. In fact, if the contractor carries out an instruction which the contract does not authorise the architect to issue, the employer is under no legal obligation to pay for the results. Indeed, the contractor would be in breach of contract. In such a case the contractor cannot take any legal action against the architect under the contract, to which the architect is not a party, but possibly the architect may find that the contractor has grounds for direct redress. Of course the contractor's chances of success will depend on all the circumstances. In view of the fact that the contractor would have no contractual remedy through the main contract, the courts might be prepared to consider an action in tort.

In order to prove negligence, three criteria must be satisfied:

(1) There must be a legal duty of care; *and*
(2) There must be a breach of that duty; *and*
(3) The breach must have caused damage.

To take an example, if the architect instructed the contractor to carry out work on land owned by the employer but outside the site boundaries as shown on the contract drawings, the architect would be acting outside any express or implied authority of the contract. The contractor would be foolish to carry out such an instruction because it would leave itself open to action by the employer for trespass at the very least. The architect would possibly be on the receiving end of actions from both contractor (in tort) and employer (under the architect's terms of engagement).

It is thought that an architect would be unwise to rely on the decision in *Pacific Associates Inc* v. *Baxter* (1988) to offer protection against legal action by the contractor in the light of other, more recent, decisions. The architect, in any event, will still be liable to third parties under the rule in *Hedley Byrne & Partners* v. *Heller & Co Ltd* (1963), for negligent statements if it can be shown that the third party acted in reliance on the advice and the architect knew of that reliance and accepted it. The architect will have liability to the client on this basis which will run together with the liability under the conditions of the engagement: *Richard Roberts Holdings Ltd* v. *Douglas Smith Stimson & Partners* (1988). Indeed, an architect's tortious liabilities may exceed those contained in the conditions of engagement: *Holt* v. *Payne Skillington* (1995). The courts have shown signs of moving the boundaries of *Hedley Byrne* liability to include duties other than statements: *Henderson* v. *Merrett Syndicates* (1994); *Conway* v. *Crowe Kelsey and Partners* (1994). Actions by contractors against architects based on reliance on negligent drawings, specifications and possibly contract administration or even tender information (*J. Jarvis & Sons Ltd* v. *Castle Wharf Developments and Others* (2001)), are becoming a possibility again.

The architect's powers and duties (see Tables 4.1 and 4.2) flow directly from the agreement with the employer. It is always prudent for them to enter into a formal written contract following the terms of the RIBA Standard Form of Agreement for the Appointment of an Architect (SFA 99) or Conditions of Engagement (CE/99) which is probably more appropriate for small works. The actual small Works version (SW/99) may be too simple for all but the smallest Works. This contract will determine the precise extent of the architect's authority, but it is important to remember that, whatever the terms of the appointment may be, it has no effect on the building contract between contractor and employer. Therefore, provided that the exercise of the architect's authority is within the limits laid down in the building contract, the contractor may safely carry out the architect's instructions, take notice of certificates, etc. without worrying whether the architect has obtained the employer's consent.

For example, the Conditions of Engagement (CE/99, April 2004 update) state (clause 2.7) that the architect must not make any material amendment to the approved design without the employer's consent, and lay an obligation upon the architect to notify the employer if the services, the fees, any other part of the appointment or any information or approvals need to be varied (clause 1.6). The contract, however,

Table 4.1
Architect's powers under MW and MWD

Clause	Power	Comment
2.1.3 (MWD only)	Direct the contractor in regard to the integration of the CDP	The directions must not affect the design of the CDP without the contractor's consent
2.10 (2.11 under MWD)	Instruct contractor not to make good defects	No express power to deduct money equal to the cost of making good but it is likely such power is derived from the architect's power to value variations under clause 3.6
3.3.1	Consent in writing to subcontracting	Consent must not be unreasonably withheld
3.4	Issue written instructions	
3.5	Require the contractor to comply with an instruction by serving written notice	The contractor has seven days from receipt of the written notice in which to comply. If it fails to do so the employer may employ and pay others to carry out the work and an appropriate deduction may be made from the contract sum
3.6	Order an addition to or omission from or other change in the Works or the order or period in which they are to be carried out or (in the case of MWD) a change in the Employer's Requirements	The variation is to be valued by the architect on a fair and reasonable basis unless its value is agreed with the contractor before the instruction is carried out
3.6	Agree the price of variations with the contractor prior to the contractor carrying out the instruction	
3.8	Issue instructions requiring the exclusion from the Works of any person employed thereon	The power must not be exercised unreasonably or vexatiously
6.4.1	Give notice to the contractor specifying a default and requiring it to be ended	The contractor has seven days in which to end the default

Table 4.2
Architect's duties under MW and MWD

Clause	Duty	Comment
2.3 (2.4 under MWD)	Issue any further information necessary for the proper carrying out of the Works and issue all certificates and confirm all instructions in writing	See Table 4.3 Clause 3.4 specifies the procedures
2.4 (2.5 under MWD)	Correct inconsistencies in or between the contract drawings, specification and schedules	If the correction results in a change it is to be valued as a variation under clause 3.6
2.7 (2.8 under MWD)	Make in writing such extension of time as may be reasonable	If it becomes apparent that the Works will not be completed by the stated completion date and the causes of delay are beyond the control of the contractor including compliance with an architect's instruction not due to the contractor's default and the contractor has so notified the architect
2.9 (2.10 under MWD)	Certify the date when the Works have reached practical completion and the contractor has complied sufficiently with clause 3.9.3	Clause 3.9.3 requires the contractor to provide and to ensure that the subcontractors provide such information as the planning supervisor reasonably requires for the health and safety file
2.10 (2.11 under MWD)	Notify the contractor of defects in the Works	The contract does not state when the notification should take place
2.11 (2.12 under MWD)	Certify the date when the contractor has discharged its obligations in respect of defects liability	The contractor's obligation is limited to remedying defects, shrinkages or other faults appearing within the defects liability period and which are due to materials or workmanship not in accordance with the contract
3.4	Confirm oral instructions in writing	This must be done within two days of *issue*

Table 4.2 *Contd*

Clause	Duty	Comment
3.6.3	Value variation instructions	The value is to be on a fair and reasonable basis and where relevant the prices in the priced documents must be used. Any direct loss and/or expense incurred by the contractor must be included if due to regular progress affected by compliance with instructions or due to compliance or non-compliance by the employer with clause 3.9
3.7	Issue instructions as to expenditure of provisional sums	The instruction is to be valued as a variation under clause 3.6.2 or 3.6.3
4.3	Certify progress payments to the contractor	Such payments are to be certified at intervals of four weeks calculated from the date of commencement
4.5	Certify payment to the contractor of the percentage stated in the contract particulars of the total amount to be paid to him	This must be done within 14 days of the date of practical completion certified under clause 2.9 (2.10 under MWD)
4.8.1	Issue a final certificate	Provided the contractor has supplied all documentation reasonably required for the computation of the final sum and the architect has issued a clause 2.11 (2.12 under MWD) certificate (defects liability). Subject to that, the final certificate must be issued within 28 days of receipt of the contractor's documentation
5.4A.2.3	Issue certificates	To pay insurance money to the contractor
5.4B.2	Issue instructions for the reinstatement and making good of loss or damage	Should loss or damage be caused by clause 5.4B events

empowers the architect (clauses 3.4 and 3.6) to issue instructions which may both alter the design and increase the total cost of the Works. The contractor must carry out such instructions and the employer is bound to pay. If the employer did not consent to the alteration in design or the increased expenditure, the employer's remedy lies against the architect.

If the architect fails to carry out any duties under the contract, it may be held to be a default for which the employer will be liable to the contractor: *Croudace Ltd* v. *London Borough of Lambeth* (1986). In *Penwith District Council* v. *V. P. Developments* (1999) it was held that the employer's duty does not arise until he or she is aware of the need to remind the architect of an obligation. The employer is not liable for what the architect does or fails to do in the capacity as certifier, because the parties have given the architect authority to form and express opinions. However, the employer is responsible for controlling the architect on becoming aware that the architect is not acting in accordance with the contract.

In performing professional duties the architect will be expected to act with the same degree of skill and care as the average competent architect: see *Chapman* v. *Walton* (1833). If the architect professes to have skills or experience which are greater than average, judgment may be stricter accordingly. Thus, if architects hold themselves out as experts in particular types of buildings, they will be expected to show a higher degree of skill in that particular area. If something goes wrong, to say that fellow architects without that specialisation would have taken the same action will be insufficient defence. It would have to be shown that a body of specialised opinion would have come to the same conclusion. The body need not be substantial in numbers provided that it exists and is responsible: *De Freitas* v. *O'Brien* (1995).

This is not the place to discuss in detail architects' general duties to their clients and to third parties. One aspect of those duties, however, does affect administration of the contract. That is the duty to the client to be familiar with those aspects of the law which affect architectural work. An architect is not expected to have the detailed knowledge of a specialist building contracts lawyer, but must be capable of advising the client regarding the most suitable contract for a particular project. This implies that the architect must have, at least, a working knowledge of the main forms of contract. In addition, the architect must be able to give detailed advice on the choice of contract. Although problems may arise during the currency of the contract which clearly demand the attentions of a legal expert, the architect's client will be less than impressed and may well consider to be incompetent an architect who is unable to explain the basic provisions of the contract and deal with the day-to-day running of the job without legal assistance. If the client is put to unnecessary expense due to the fact that the architect had inadequate knowledge of the contractual provisions, the client may well sue. The architect is also expected to be aware of decisions of the courts relevant to architecture and construction: *B. L. Holdings* v. *Robert J. Wood & Partners* (1979).

Up to the moment the employer and contractor enter into the contract, the architect has been acting as agent with limited authority for the employer. During the contract the architect is expected to continue to act as the employer's agent, but also to administer the terms of the contract fairly between the parties. The employer may find the change difficult to understand and, to avoid problems, the wise architect will explain the situation, perhaps at the same time as sending the contract documents for signature (Figure 4.1). The architect is not in a quasi-arbitral role and, therefore, not immune from any action for negligence by either party. The point was established by *Sutcliffe* v. *Thackrah* (1974).

The result is that the architect may be in the position of deciding his or her own default, e.g. in considering an extension of time. In the case of *Michael Sallis & Co Ltd* v. *Calil and W. F. Newman & Associates* (1987), it was held that an architect owes a duty of care to a contractor when the contract calls on the architect to act fairly between the parties. Such things as extensions of time and certification of money would fall under this heading. The decision was called into question by the Court of Appeal in *Pacific Associates Inc* v. *Baxter* (1988), a case which concerned a firm of consulting engineers. The grounds for the *Pacific Associates* decision involved the existence of an arbitration clause in the main contract and an exclusion of liability clause. The decision effectively reduced the contractor's choice of parties against whom to take action. The existence of an exclusion of liability clause is not relevant to the question of whether a duty of care exists. The arguments put forward in *Sallis* may yet be revived in a higher court particularly in the light of more recent decisions noted above.

If the employer suffers loss or damage due to the fact that the architect has issued information late and the architect has granted an extension on that basis, the employer may well seek to recover that loss from the architect. It might be thought that the architect's position is slightly less hazardous under this contract than it is under SBC or IC because of the absence of a full provision enabling the contractor to claim 'loss' and 'expense' under the contract as a result of architect's defaults, e.g. through late supply of information. However, if the contractor brought a successful action against the employer at common law, the employer

Figure 4.1
Letter from architect to employer explaining the duty to act impartially

Dear Sir

PROJECT TITLE

[*Insert the main point of the letter and then continue:*]

This is probably an opportune moment to explain the nature of the additional responsibility which I carry during the currency of the contract. Until the contract is signed I am required, in accordance with the conditions of my appointment, to act solely as your agent within the limits laid down in those conditions. Thereafter, although I continue to act for you as before, I have the additional duty of administering the contract conditions fairly between the parties. In effect, this means that I must make any decisions under the contract strictly in accordance with the terms of the contract. If you require any further explanation of the position, I would be delighted to meet you for that purpose.

Yours faithfully

might well be able to recoup the loss from the architect. Even if no such action were brought, the employer might sue the architect for loss of liquidated damages.

The extent of the authority and duties of an employee of the employer, for example in local or central government, is not always very clear. If the architect is in this position, the contractor may well consider that the architect at certain times is acting as agent for the employer and that any instruction issued which is not expressly empowered by the contract is, in effect, a direct instruction from the employer. The dangers of incurring unexpected costs are obvious and the architect should ensure that the extent of authority is made clear to the contractor at the beginning of the contract (Figure 4.2).

As architect employee, the duty to act fairly between the parties remains, but it is admittedly difficult for the architect to convince the contractor that he or she is acting impartially. The administration of the contract under these circumstances calls for complete integrity, not only on the part of the architect but also on the part of the employer who has a duty to ensure that the architect carries out professional duties properly in accordance with the contract: see *Perini Corporation* v. *Commonwealth of Australia* (1969). Although it is usual for the actions of local authority employees to be governed by Standing Orders of the Council, they are of no concern to the contractor unless they have been specifically drawn to its attention at the time of tender. Thus an order that only chief officers may sign financial certificates would not avail the architect as a defence if the architect was unable to issue a certificate because the chief officer was unavailable. The contractor is entitled to rely on the architect's apparent authority. Orders that certificates may not be issued without authority from the audit department are likewise irrelevant so far as the contractor is concerned.

The best summary of the legal situation is this:

> 'An architect is usually and for the most part a specialist exercising his special skills independently of his employer. If he is in breach of his professional duties he may be sued personally. There may, however, be instances where the exercise of his professional duties is sufficiently linked to the conduct and attitude of the employers so as to make them liable for his default.' (Kilner Brown J in *Rees & Kirby Ltd* v. *Swansea Corporation* (1983))

Although this case subsequently went to appeal, it is suggested that Kilner Brown J's statement remains an accurate expression of the law.

Figure 4.2
Letter from architect to contractor if architect is employee in local authority, etc.

Dear Sir

PROJECT TITLE

Possession of the site will be given to you by the employer on the [*insert date*] in accordance with the contract provisions.

In order to avoid any misunderstandings which might arise in the future I should make clear, as far as you are concerned, that the extent of my authority is laid down in the contract. Although I am an employee of [*insert name of employing body*], I have no general power of agency to bind the employer outside the express contract terms. If I have cause to write to you on behalf of the employer, I will clearly so state. If any matters fall to be decided by me under the contract, I will make such decisions impartially between the parties.

Yours faithfully

4.2 Express provisions of the contract

Article 3 of MW provides for the insertion of the name of the architect. The person so named will be the person referred to in the conditions whenever the word 'architect' appears. It is common practice to insert the name of the architectural firm (or the name of the chief architect in a local authority) because of the difficulties which could arise in having to renominate every time the project architect left the firm, died or retired. It is seldom the cause of any dispute, but it is prudent to notify all interested parties of the name of the authorised representative, i.e. the project architect, who will administer the contract on a day-to-day basis (Figure 4.3). Changes in the identity of the project architect should be notified in the same way. The case of *Croudace Ltd* v. *London Borough of Lambeth* (1986), establishes that at common law the employer is under a duty to appoint another architect if, for example, the named architect resigns, retires or dies, and Article 3 makes this duty explicit.

It is not strictly necessary for the architect named in Article 3 to sign all letters, notices, certificates and instructions personally, but they must be signed by a person duly authorised. It is not wise for the authorised representative to sign his or her name only, even if using headed stationery, because:

- The letter, certificate, etc. must be signed by or in the name of the architect named in the contract: *London County Council* v. *Vitamins Ltd* (1955).
- Otherwise the letter may be deemed to be written merely on the authority of the person who signs, with serious financial consequences.

The architect should not:

- Use a rubber stamp. It is not good practice although it is probably valid.
- Sign someone else's name and add initials. To sign in someone's name is best accomplished by procuration ('per pro' or 'pp').

If the named architect dies or ceases to act for some other reason, Article 3 states that it is the employer's duty to nominate a successor within 14 days of the death or ceasing to act of the named architect: *Croudace Ltd* v. *London Borough of Lambeth* (1986). Unlike the position under SBC, there is no provision for the contractor to object to the successor, but there is nothing to prevent the contractor from referring the matter to

Figure 4.3
Letter from architect to contractor naming authorised representatives

Dear Sir

PROJECT TITLE

This is to inform you formally that the architect's authorised representatives for all the purposes of the contract are:

[*insert name*] – Partner in charge of the contract.
[*insert name*] – Project architect.

Until further notice the above are the only people authorised to exercise the authority of the architect in connection with the contract.

Yours faithfully

for and on behalf of [*insert the name of the architect/practice in the contract*]

Copies: Employer
Quantity surveyor
Consultants
Clerk of works

immediate adjudication or arbitration if it feels strongly about it. The contractor must, however, continue with the works while awaiting the outcome of the dispute resolution process and, in the majority of contracts carried out under this form, it is likely that the Works will be complete by the time a decision is reached. That is probably the reason for the exclusion of the right of objection in the first place. Clearly, the employer would do well to listen carefully if the contractor makes any representations, in order to avoid such difficulties. It is regrettable that an employer will occasionally try to take over the former architect's role personally. An employer is not entitled to do that, but must appoint another architect to the job: *Scheidebouw BV* v. *St James Homes (Grosvenor Dock) Ltd* (2006).

There is an important proviso to this Article which states that 'no Architect ... appointed for this Contract shall be entitled to disregard or overrule any certificate, opinion, decision, approval or instruction given by' the former architect unless the former architect would have had the power to do so. This is a sensible provision to safeguard the contractor's interests under circumstances which are in the sole control of the employer. A successor architect who disagrees with some of the predecessor's decisions should notify the employer immediately (Figure 4.4), to safeguard the position. The proviso clearly cannot mean that certificates or instructions given by the former architect must stand for all time, even if they are wrong. The implication of the proviso is that if it is necessary for the successor architect to make changes, they will be treated as variations and the contractor will be entitled to payment accordingly. For example, if the former architect had given instructions for the construction of a detail which, in the opinion of the new architect, would lead to trouble, the new architect could issue further instructions correcting the matter and the contractor would be paid for correcting the work. Unlike instructions, certificates are expressions of the architect's opinion in tangible form for the purposes specified in the contract: *Token Construction Co Ltd* v. *Charlton Estates Ltd* (1973). Certificates of practical completion and making good are not, in any event, susceptible to change. On the other hand, an architect is normally entitled to amend a financial certificate at the next time for payment and, being cumulative, it is difficult to see how a new architect could forfeit this right.

Although clause 2.1 is entitled 'Contractor's obligations,' it creates a problem for the architect. After concisely stating the contractor's obligation to carry out and complete the works in a proper and workmanlike manner in accordance with the contract documents, the Health

Figure 4.4
Letter from architect to employer if disagreeing with former architect's decisions, etc.

Dear Sir

PROJECT TITLE

I have now had the opportunity of examining all the drawings, files and other papers relating to this contract and I have visited the site and spoken to the contractor.

Article 3 of the contract prevents me from disregarding or overruling any certificate or instruction given by the architect previously engaged to administer this contract. This does not prevent me from issuing further certificates and, particularly, further instructions which may vary instructions given by the former architect. Of course in the latter instance there will be a cost implication.

I list below the matters on which I find myself unable to agree entirely with the decisions of the former architect:

[*list all matters which which you disagree*]

When you have had the opportunity to study these matters I suggest we should meet to discuss ways of dealing with them.

Yours faithfully

and Safety Plan and statutory requirements there is a proviso in clause 2.1.2:

> 'Where and to the extent that approval of the quality of materials or of the standards of workmanship is a matter for the opinion of the architect... such quality and standards shall be to his reasonable satisfaction...'.

It has been held that such a proviso in the JCT 80 and IFC 84 forms of contract were to be given a very wide meaning: *Crown Estates Commissioners* v. *John Mowlem & Co Ltd* (1994); *Colbart Ltd* v. *H. Kumar* (1992). In effect, the quality and standards of all materials and workmanship are matters for the architect's reasonable satisfaction. If the contractor disagrees that the architect is being reasonable, it can, of course, seek adjudication on the point. The danger is threefold:

- It might be argued that it is implicit in the proviso that the architect will in due course and especially if requested by the contractor, express satisfaction about all matters. The contract provides no machinery for so doing, but the wise contractor will ask the architect to express the satisfaction in writing (see Figure 4.5) and it is difficult to see how the architect could be justified in refusing unless dissatisfied.
- The final certificate is not stated to be conclusive about anything, thus avoiding the trap which used to exist in JCT 80 and IFC 84, but it would not be easy for the architect to justify issuing the final certificate and becoming *functus officio* if there are matters with which the architect is dissatisfied.
- If the architect formally approves quality and standards, this approval will override any requirements in the contract documents, so if the architect approves something which is not in accordance with the contract, the employer is prevented from seeking redress from the contractor on the ground of lack of compliance.

Clause 2.3 very briefly sets out the architect's duties. These are:

- The issue of further information
- The issue of all certificates in accordance with the contract (see Table 4.3)
- The confirmation of all instructions in writing in accordance with the contract.

Figure 4.5
Letter from contractor to architect seeking approval

Dear Sir

PROJECT TITLE

We refer to item [*insert item number*] which states that [*specify material or work*] is to be to your approval. Clause 2.1.2 of the conditions of contract state that in such instances the quality and standards are to be to your reasonable satisfaction.

The above mentioned material has been installed/work has been completed [*delete as appropriate*] and we should be pleased to receive your written approval so as to comply with the terms of the contract.

Yours faithfully

Table 4.3
Certificates to be issued by the architect under MW and MWD

Clause	Certificate
2.9 (2.10 under MWD)	Practical completion
2.11 (2.12 under MWD)	Making good
4.3	Progress payments
4.5	Penultimate certificate
4.8.1	Final certificate
5.4A.2.3	Insurance payments

The amount, if any, of further information necessary will vary greatly depending on the size and complexity of the work. In the case of small simple projects, the contract drawings may need very little elaboration. Whether the information is necessary should be a matter of fact, not opinion. The contractor is expected to use its own practical experience in carrying out the work: *Bowmer & Kirkland Ltd* v. *Wilson Bowden Properties Ltd* (1996). Note that there is no obligation on the contractor to apply for further information either in this clause or anywhere else in the contract. The prudent contractor will do so, but it is the architect's duty to supply correct information to enable the contractor to carry out the Works properly in accordance with the contract: *London Borough of Merton* v. *Stanley Hugh Leach Ltd* (1985). The completion date is part of the provisions of the contract, therefore the architect's duty is to supply necessary information at the correct times to enable the contractor to proceed with the works and to complete them by the contract completion date. If the architect fails to do so, it will be a breach for which the contractor can claim extension of time and possibly damages at common law. Of course, a contractor who merely stands by while the architect fails to issue information at the right time is unlikely to attract much sympathy from an adjudicator or an arbitrator. It is always wise for a contractor to ask for information it requires, allowing a reasonable period for the architect to respond.

The issue of certificates refers not only to progress payments but also to such things as practical completion and making good after the rectification period. Although there is no prescribed way in which a certificate should be set out, it is a formal document. It may be in the form of a letter, but for the avoidance of doubt it is always good practice to head the letter 'Certificate of . . . ' and begin 'I certify . . . '. Where a number of certificates are to be issued in sequence it is normal to number them in order. Delay in the issue of a certificate is a serious matter and may give rise to financial claims by the contractor (Table 10.1).

Confirming instructions will be dealt with in section 4.3.

The remaining express duties of the architect are covered in the appropriate chapters of this book. Because this contract is so brief, the duties are also briefly stated. This can be misleading and the architect should always bear in mind that the courts will imply terms to cover the way in which the architect must administer the contract. Generally, the architect is expected to act promptly and efficiently. An architect who does so will avoid claims and contribute towards the smooth running of the contract.

4.3 *Architect's instructions*

All instructions to the contractor must be confirmed in writing to be effective. Despite what some contractors maintain, there is absolutely no requirement that instructions must be on a specially printed form headed 'Architect's Instruction', although it is undoubtedly good practice to use such forms because:

- It leaves no room for doubt that the architect is issuing an instruction. An instruction buried in a letter which deals with a great many other things can sometimes be confusing.
- It makes the job of keeping track of instructions for the purposes of valuation and checking so much easier.

Instructions can be given in letter form, provided it is made clear that the architect is issuing an instruction (Figure 4.6). Hand-written instructions can also be given on site, provided they are signed and dated (the architect should always sign on behalf or in the name of the architect named in the Articles). Instructions contained in the minutes of site meetings are valid if the architect is the author of the minutes and if they are recorded as agreed at a subsequent meeting. It is probable, however, that such instructions are not effective until the contractor receives a copy of the minutes' recording agreement. Since site meetings are sometimes at monthly intervals, the matter giving rise to the instruction may be ancient history before the contractor's duty to comply becomes operative; minuted instructions are, therefore, best avoided.

The position with regard to drawings is uncertain. If the architect issues a drawing together with a letter instructing the contractor to use the drawing for the Works, that is certainly an instruction for contract purposes. If the drawing is simply issued under cover of a compliments slip, the drawing may be an instruction or it may simply be sent for comment; in most cases it will be taken to be an instruction. The contractor would be wise to check first, but the architect may have difficulty showing that the drawing was not intended as an instruction if the contractor simply carries out the work. Note that if the architect hands the contractor a copy of the employer's letter requiring some action under cover of a compliments slip, it may not be an instruction at all. The moral is clear: compliments slips should not be used where an instruction is intended.

Clause 3.4 is the principal contract provision governing the issue of instructions. The procedure is shown in Figure 4.7. Despite what, at first sight, may appear to be an all-embracing provision – 'The

Figure 4.6
Letter from architect to contractor issuing an instruction

Dear Sir

PROJECT TITLE

In accordance with clause 2.10/3.4/3.6/3.7/3.8/5.4B.2 [*delete as appropriate*] please carry out the following instructions forthwith:

[*insert the instruction*]

Yours faithfully

Figure 4.7
Flowchart of architect's instructions

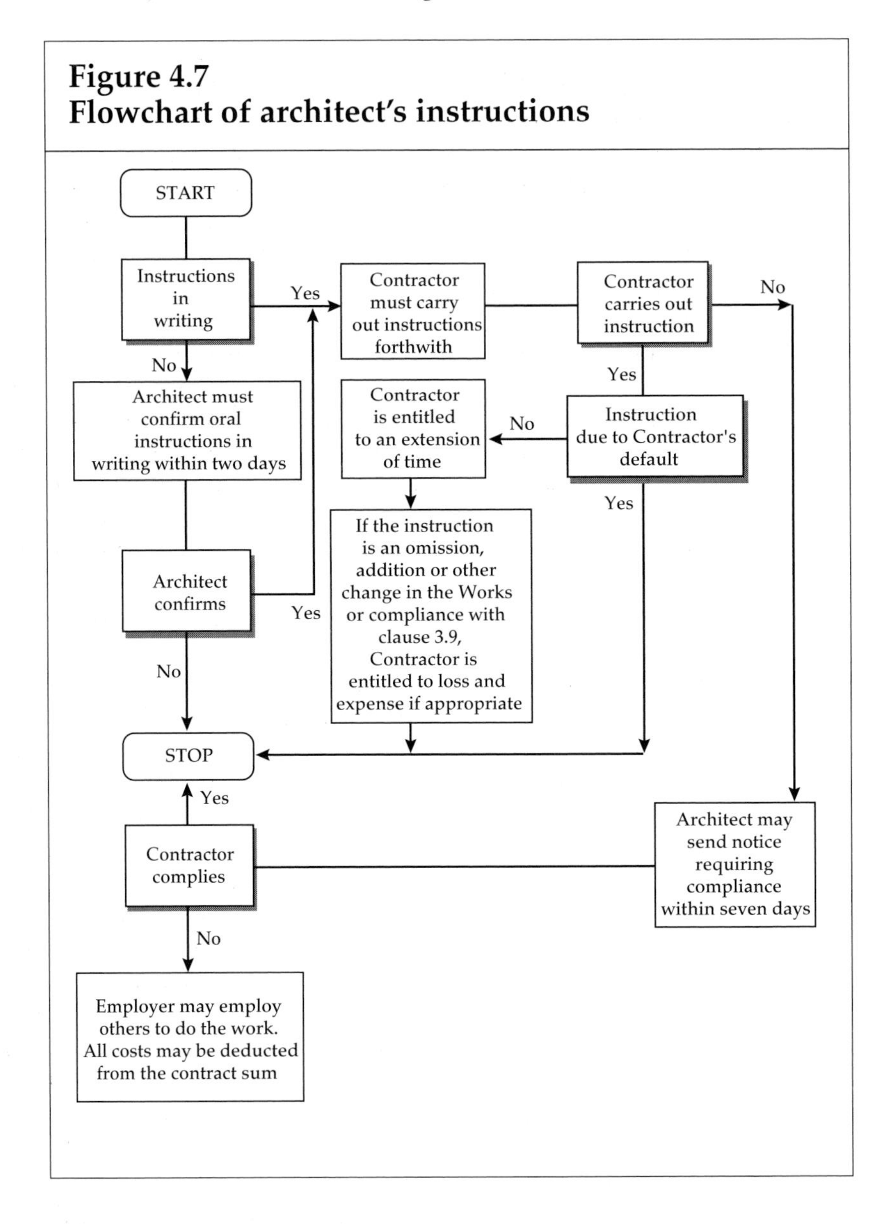

Architect... may issue written instructions which the Contractor shall forthwith comply with' – the architect must act within the scope of authority and the instruction must relate to the contract Works. The courts do not generally favour sweeping generalisations and prefer to see respective rights and duties expressed in precise terms. In the event of a dispute, therefore, it is probable that the clause will be given a very narrow interpretation. The contractor must carry out the instruction 'forthwith', i.e. as soon as it reasonably can: *London Borough of Hillingdon* v. *Cutler* (1967).

Provision is made for the situation if instructions are issued orally. They must be confirmed in writing by the architect within two days. The contract is silent as to when such an instruction becomes effective. It is considered that, in context, the instruction would not become effective until the contractor receives the confirmation. Although the practice of issuing oral instructions is widespread, it is difficult to see why it was necessary to make special provision in the contract since the effect of the confirmation is no different from the effect of a simple written instruction in the first place. The oral instruction itself is irrelevant. Note that there is no provision for the contractor to confirm oral instructions. But if the contractor confirms an oral instruction, the architect remains silent and the contractor proceeds with the work, it is likely that the employer will be estopped (prevented) from denying the contractor's entitlement to payment: *Bowmer & Kirkland Ltd* v. *Wilson Bowden Properties Ltd* (1996). *Redheugh Construction Ltd* v. *Coyne Contracting Ltd and British Columbia Building Corporation* (1997) and *Ministry of Defence* v. *Scott Wilson Kirkpatrick and Dean & Dyball Construction* (2000) reach broadly similar conclusions on this point. However, a contrary view was taken in *W. S. Harvey (Decorators) Ltd* v. *H. L. Smith Construction Ltd* (1997). Ideally, oral instructions should be avoided.

If the contractor does not carry out instructions 'forthwith', the contract provides a remedy in much the same form as SBC, clause 3.11. As a first step, under clause 3.5 the architect must send the contractor a written notice requiring him to comply with the instruction within seven days (Figure 4.8). If the contractor does not so comply, the architect should advise the employer (Figure 4.9) that others may be employed to carry out the work detailed in the instruction. Although the contract specifically states that it is the employer who may employ others, the employer will expect the architect to give advice and handle the details. It will amount to a completely separate contract and, in order to ensure that there can be no reasonable question about costs to which the employer is entitled, the architect should obtain competitive quotations from three firms if time and circumstances permit: *Rhuddlan Borough*

Figure 4.8
Letter from architect to contractor requiring compliance with instructions before default action taken

RECORDED DELIVERY/SPECIAL DELIVERY*

Dear Sir

PROJECT TITLE

On [*insert date*] I instructed you to [*specify*] under clause 3.4 of the contract. That clause requires you to carry out my instructions forthwith.

You have failed to do so and in accordance with clause 3.5, I hereby require you to carry out the above-mentioned instruction. Should you fail to comply within seven days after receipt of this notice, the employer will engage others to carry out the work and all the resultant costs will be deducted from the contract sum.

Yours faithfully

Copies: Employer
Quantity surveyor (if appointed)

* *Service by recorded delivery or special delivery is not required by clause 3.5 but is a desirable precaution.*

Figure 4.9
Letter from architect to employer if contractor fails to comply with instruction within seven days after notice

Dear Sir

PROJECT TITLE

I refer to my letter to [*insert name of contractor*], dated [*insert date*], requiring him to comply with my instruction of the [*insert date*] within seven days.

The seven days expired yesterday and, during a site visit this morning, I observed that the contractor had not complied with my instruction.

I advise that you may take advantage of the remedy afforded by clause 3.5 of the Conditions of Contract and, if you will let me have your written instructions to that effect, I will obtain competitive tenders from other firms for the carrying out of the work. All additional costs, including my additional fees, will be deducted from the contract sum.

Yours faithfully

Council v. *Fairclough Building Ltd* (1985). After the work is completed, the architect may make an appropriate deduction from the contract sum in this regard.

Note that the 'additional costs' refer to additional costs over and above the cost of the instruction. It is not intended that the employer should get the work done at the contractor's expense.

The additional costs may include such items as scaffolding, cutting out and making good, depending on circumstances. The architect is usually entitled to charge additional fees and the architect may include them in the computation of costs together with any other incidental expenses attributable to the contractor's non-compliance. When preparing the final account, the contractor is entitled to a brief statement showing how the figure has been calculated.

Note that there is no provision for the contractor to object to any instruction, or request the architect to state the empowering provision. However, Figure 4.10 is the kind of letter the contractor may send when receiving a compliance notice and the contractor can always seek immediate adjudication on the matter. Apart from clause 3.4, there are only five other clauses empowering the architect to issue instructions in particular circumstances:

(1) Clause 2.10 empowers the architect to instruct the contractor not to make good defects at its own cost (this provision is discussed in detail in Chapter 9).
(2) Clause 3.6 empowers the ordering of variations by way of additions, omissions or other changes in the Works or the order or period in which they are to be carried out.
(3) Clause 3.7 requires the architect to issue instructions regarding the expenditure of any provisional sum (see Chapter 11).
(4) Clause 3.8 empowers the architect to issue instructions requiring the exclusion from the Works of any person employed thereon. The instruction is not to be issued unreasonably or vexatiously and is clearly intended to enable the architect to have incompetent operatives removed.
(5) Clause 5.4B.2 requires the architect to issue instructions for the reinstatement and making good of loss or damage caused by fire, lightning, storm, etc. (see Chapter 3).

Clause 3.2 provides that any instructions given to the contractor's representative on the Works are deemed to have been issued to the contractor. In the context of the contract as a whole, such instructions

Figure 4.10
Letter from contractor to architect on receipt of seven-day notice requiring compliance with instruction

Dear Sir

PROJECT TITLE

We have today received your notice dated [*insert date*] which you purport to issue under clause 3.5 of the above contract.

[*Add either:*]

We will comply with your instruction Number [*insert number*] dated [*insert date*] forthwith, but our compliance is without prejudice to and reserving any other rights or remedies which we may possess.

[*Or:*]

We consider that we have already complied with your instruction Number [*insert number*] dated [*insert date*]. If the employer attempts to employ other persons and/or if you deduct any monies from the contract sum this will constitute a serious breach of contract for which we shall seek appropriate remedies. However, without prejudice to the foregoing, provided you immediately withdraw your purported notice of non-compliance, our [*insert name*] will be happy to meet you on site or at your office to sort out what appears to be an unfortunate misunderstanding.

[*Or:*]

It is not reasonably practicable for us to comply within the period specified by you because [*give reasons*]. You may be assured that we are well aware of our contractual obligations and intend to carry out your instruction Number [*insert number*] dated [*insert date*] as soon as [*indicate operation, etc.*] is complete [*or as appropriate*]. In light of this explanation, perhaps you will be good enough immediately to withdraw your notice requiring compliance.

Yours faithfully

Copy: Employer

must be in writing for the clause to have any effect, although in practice the architect seldom gives written instructions to the person in charge. Site instructions tend to be oral and are therefore ineffective, written instructions being sent to the contractor's main office.

It has already been mentioned that the courts will probably construe clause 3.4 narrowly as regards the architect's powers to issue instructions. Although not expressly stated, it is considered that the following common situations will fall within the architect's powers under the clause:

- The correction of inconsistencies between the contract documents. Although clause 2.4 deals with this matter, it does not specifically provide that the architect can issue instructions. Nonetheless, it is considered that the architect certainly has that power.
- Opening up of work for inspection and testing of materials. If the work or materials are found to be in accordance with the contract, the contractor will have a claim for extension of time. It is thought that, under those circumstances, the costs of carrying out the instruction could be valued under clause 3.6, but if the contractor considers that it has been involved in loss and/or expense, that would have to be the subject of a common law claim (see Chapter 10).
- The removal or correction of defective work. The clause appears to be broad enough to cover either requirement.
- Probably the postponement of any work in progress. This power could not extend to deferring the giving of possession of the site which would amount to varying an express term of the contract – a matter for the parties to negotiate. Any postponement instruction would inevitably give rise to an extension of time. It was previously thought that postponement would inevitably give the contractor an entitlement to damages at common law. On reflection, the position is probably not so simple. There would certainly be no express entitlement to loss and/or expense under clause 3.6, but because no authorised instruction is a breach of contract, there would be no entitlement to damages for breach of contract either. The conclusion is that if the architect has power to postpone work as an implied term, there must be a further implied term entitling the contractor to further reimbursement. Certain categories of postponement concerning only a part of the Works could conceivably be brought under the terms of clause 3.6: 'The Architect... may issue instructions requiring... change in the Works or the order or period in which they are to be carried out...', so enabling the architect to include direct loss and/or expense in the valuation.

4.4 Summary

Authority

- From the contractor's point of view, the architect's authority is defined by the terms of the contract.
- The contractor can only take action against the architect in tort. The chances of success appear to be improving.
- The employer can take action against the architect in contract and in tort.
- The architect's authority depends on what has been agreed with the employer.
- The architect is expected to use the same degree of skill as any other average, competent architect unless professing a higher degree of skill in a particular area.
- Once the contract is in effect, the architect must act fairly to both parties.
- The architect does not have a quasi-arbitral role nor any immunity from actions for negligence as a result of the architect's decisions.
- Architects who are employees of the employer may be in a difficult position.
- The architect cannot overrule the certificates or instructions of a previous architect engaged on the work.
- The architect's approval will generally override contract requirements.
- All instructions must be confirmed in writing by the architect.
- The architect has implied as well as express duties.
- The architect's general power to issue instructions is confined to matters relating to the Works.

CHAPTER FIVE
CONTRACTOR

5.1 *Contractor's obligations: express and implied*

5.1.1 Legal principles

It is all too commonly assumed that the whole of the contractor's contractual obligations are contained and set out in the printed contract form. This is not so and this particular contract form is silent on many important matters. These gaps are filled in by terms which the law will write into the contract.

If nothing at all were said about the contractor's obligations, the law would require the contractor to do three things:

(1) To carry out its work in a good and workmanlike manner exercising reasonable care and skill. This means that the contractor must show the same degree of competence as the average contractor experienced in carrying out that type of work.
(2) To use materials of good quality which are reasonably fit for their purpose.
(3) To ensure that the completed building or structure is reasonably fit for its intended purpose provided that purpose is known. This obligation is modified where the employer engages an architect to design the Works since the architect is then responsible for the design.

Similarly, a term would be implied that the contractor would comply with the building regulations and with other statutory requirements – a matter which is the subject of an express term in MW.

These implied terms can be modified by the express terms of the contract itself and are, in fact, modified by what MW says, with the result that the contractor is under a lesser duty than would otherwise be the case.

However, statute also imposes similar implied obligations on the contractor and in particular the Supply of Goods and Services Act 1982 implies terms as to the quality of goods supplied by the contractor under the contract. The construction of dwellings (both houses and flats) is governed by the Defective Premises Act 1972 (in Northern Ireland, the Defective Premises (Northern Ireland) Order 1975) which provides:

> 'Any person taking on work for or in connection with the provision of a dwelling... owes a duty to see that the work that he takes on is done in a workmanlike manner, with proper materials... and so as to be fit for the purpose required...'.

The Act envisages a two-stage test. Therefore, if the dwelling is fit for habitation despite the fact that some work was not done in a workmanlike manner nor with proper materials, it seems that it complies with the Act: *M. C. Thompson* v. *Clive Alexander & Partners* (1992).

The express provisions of MW must, therefore, be read against this background. There is a further important point. As discussed in Chapter 8, clause 3.3.1 envisages that the contractor may sublet the works or any part of them if the architect gives written consent. Subcontracting does not free the contractor from responsibility for the subcontracted work. The contractor remains responsible to the employer for all defaults of the subcontractor as regards workmanship, materials or otherwise.

Tables 5.1 and 5.2 summarise the contractor's powers and duties respectively under the express provisions of MW.

5.1.2 Execution of the works

Clause 2.1 requires the contractor to carry out and complete the work in accordance with a number of criteria:

- in a proper and workmanlike manner
- in compliance with the contract documents
- in compliance with the Health and Safety Plan (if it applies) *and*
- in accordance with statutory requirements.

Users of MW 98 will immediately notice that the requirement to proceed with 'due diligence' has been omitted. That point is discussed below.

This is probably the most important clause in the contract. It sets out the contractor's basic obligation. The contractor is bound to complete the Works by the date for completion set out in the contract particulars. The Works must be brought to a state where they are practically completed so that the architect can issue the certificate under clause 2.9 (2.10 under MWD).

Table 5.1
Contractor's powers under MW and MWD

Clause	Power	Comment
2.2 (2.3 under MWD)	Commence the works on the specified date	
3.1	Assign the contract	If the employer gives written consent
3.3.1	Sublet the works or part thereof	Only if the architect consents in writing. Consent must not be unreasonably withheld
3.6.2	Agree the price of variations	Before executing the variation instruction
4.7	Suspend performance of obligations under the contract	If employer fails to pay, and has not issued effective notice of withholding or deduction of money, by final date for payment Contractor must first give seven days' notice and must resume work after payment in full
5.5	Reasonably require evidence from the employer that the insurance referred to in clause 5.4B has been taken out and is in force	Except where the employer is a local authority
6.8.1	Give notice to the employer specifying the default and requiring it to be ended within seven days	The notice must be served on the employer by special or recorded delivery or actual delivery Notice can be served: • If the employer fails to pay by the final date for payment the amount properly due *or* • If the employer interferes with or obstructs the issue of any certificate *or* • If the employer fails to comply with the requirements of the CDM Regulations

Table 5.1 *Contd*

Clause	Power	Comment
6.8.2	Give notice to the employer specifying the suspension event and requiring it to be ended within seven days	The notice must be served on the employer by special, recorded or actual delivery Notice may be served if the whole or substantially the whole of the Works is suspended for one month due to: • architect's instructions; *or* • impediment, prevention or default by the employer, architect, etc.
6.8.3	Terminate the contractor's employment	If default or suspension not ended after notice Termination takes effect on receipt of this notice The notice must not be given unreasonably or vexatiously
6.9	Terminate the contractor's employment	By notice if the employer is insolvent as defined in clause 6.1 The notice does not require a prior default notice and it takes effect on receipt by the employer
6.10	Give notice to the employer that unless the suspension ends within seven days, the contractor's employment may be terminated	The notice must be served on the employer by special, recorded or actual delivery Notice may be served if the whole or substantially the whole of the Works is suspended for one month due to: • *force majeure; or* • architect's instructions as a result of negligence or default of statutory authority; *or* • loss by specified perils; *or* • civil commotion or terrorism; *or* • exercise of UK government power
	Terminate the contractor's employment under the contract	By notice served on the employer by special, recorded or actual delivery if the suspension does not end

Table 5.1 *Contd*

Clause	Power	Comment
7.1	Agree to resolve the dispute by mediation	
Schedule 1 para 1	By written notice jointly with the employer to the arbitrator stating that they wish the arbitration to be conducted in accordance with any amendments to the JCT 2005 CIMAR	
Schedule 1 para 2.1	Serve on the employer a written notice	If the contractor wishes a dispute to be resolved by arbitration
Schedule 1 para 2.3	Give a further arbitration notice to the employer referring to any other dispute	After the arbitrator has been appointed Rule 3.3 applies

Table 5.2
Contractor's duties under MW and MWD

Clause	Duty	Comment
2.1	Carry out and complete the Works in accordance with the contract documents in a good and workmanlike manner and in accordance with statutory requirements and give all notices	In accordance with the Health and Safety Plan of the principal contractor if applicable
2.1.1 (MWD only)	Complete the design for the CDP using reasonable skill and care	The contractor is not responsible for the Employer's Requirements or the adequacy of any design therein
2.1.3 (2.2.2 under MWD)	Take all reasonable steps to encourage employees and agents, including subcontractors, to be registered cardholders under the Construction Skills Certification Scheme	
2.1.3 (MWD only)	Comply with architect's directions for integration of CDP work	Subject to clause 3.4.2
2.1.4 (MWD only)	Comply with CDM Regulation 13, including co-operation with planning supervisor	
2.1.5 (MWD only)	Provide the architect with two copies of drawings, etc. to explain the CDP	As and when necessary
2.2 (2.3 under MWD)	Complete the Works by the specified date	The architect must insert the date in the contract particulars
2.5.1 (2.6.1 under MWD)	Immediately give to the architect a written notice specifying the divergence	If the contractor finds a divergence between statutory requirements and the contract documents or architect's instructions

Table 5.2 *Contd*

Clause	Duty	Comment
2.5.2 (MWD only)	Correct any inconsistency in or between the CDP documents	At the contractor's own expense
2.6 (2.7 under MWD)	Pay all fees and charges in respect of the Works	Provided they are legally recoverable
2.7 (2.8 under MWD)	Notify the architect of delay	If: • It becomes apparent that the Works will not be completed by the specified date *and* • This is because of reasons beyond the control of the contractor including compliance with any instruction of the architect whose issue is not due to a default of the contractor
2.8.1 (2.9.1 under MWD)	Pay liquidated damages to the employer	If the Works are not completed by the specified date or by any later date fixed under clause 2.7 (2.8 under MWD)
2.10 (2.11 under MWD)	Make good at own cost any defects, shrinkages or other faults	The defects, shrinkages or other faults must have appeared within the specified period of the date of practical completion *and* must be due to materials or workmanship not in accordance with the contract *and* unless the architect has instructed otherwise
3.2	Keep a competent person in charge upon the Works at all reasonable times	Under clause 3.8 the architect may instruct this person's exclusion
3.4	Carry out all architect's instructions forthwith	The instructions must be in writing
3.9	Comply with all the duties of a principal contractor set out in the CDM Regulations and in this clause	Where the contractor is and remains a principal contractor

Table 5.2 *Contd*

Clause	Duty	Comment
3.9.2	Ensure that the health and safety plan is received by the employer before construction work starts Notify any amendments	If the contractor is and remains the principal contractor The employer must notify the planning supervisor and the architect, where relevant
3.9.3	Provide information reasonably required for the preparation of the health and safety file as required by the CDM Regulations and ensure any subcontractor does the same	Within the time reasonably required in writing by the planning supervisor. If the contractor is not the principal contractor, the information must be provided to the principal contractor
3.10	Comply with all the reasonable requirements of the principal contractor	At no cost to the employer where the employer has appointed a successor to the contractor as principal contractor to the extent necessary for compliance with CDM Regulations. No extension of time is to be given
4.8.1	Supply the architect with all documentation reasonably required to enable the final sum to be computed	• Within the specified period of the date of practical completion • It is not conditional upon a request from the architect
4.9	Pay simple interest to the employer on amount not properly paid in the final certificate, until paid	If the contractor fails to pay any amount due to be paid by the final date for payment
5.1	Indemnify the employer against any expense, liability, loss, claim or proceedings in respect of personal injury or death	If the expense, etc. arises out of or in the course of or is caused by reason of the carrying out of the Works except to the extent that it is due to any act or neglect of the employer or those for whom the employer is legally responsible

Table 5.2 *Contd*

Clause	Duty	Comment
5.2	Indemnify the employer against and insure and cause its subcontractors to insure against any expense, liability, loss, claim or proceedings for damage to property other than the Works	This is subject to clause 5.4B The indemnity operates if the expense etc: • Arises out of or in the course of or is caused by reason of the carrying out of the Works *and* • To the extent that it is due to any negligence, omission or default of the contractor or any subcontractor or any person for whom they are legally responsible
5.3	Take out and maintain and cause its subcontractors to maintain the insurances necessary to meet its liability under clauses 5.1 and 5.2, including compliance with all relevant legislation	
5.4A.1	Insure against the specified risks	New Works only
5.4A.2.1	Restore or replace work or materials etc., dispose of debris, and proceed with and complete the Works	After any inspection required by the insurers following a clause 5.4A claim
5.4A.2.2	Authorise payment of insurance money to employer	Employer may retain amount for professional fees
5.5	Produce evidence of insurances and cause subcontractors so to do	If the employer reasonably so requires
6.7.1	Immediately cease to occupy the site	Where the employer terminates the contractor's employment

Table 5.2 *Contd*

Clause	Duty	Comment
6.11.2	Prepare an account setting out: • Total value of work properly executed and materials properly brought on site and other amounts due *and* • Cost to the contractor of moving from site *and* • Direct loss and/or damage	Upon termination of the employment of the contractor under clauses 6.8 to 6.10. The employer must pay the full amount properly due within 28 days of submission of the account

Although the contractor's basic obligation is not qualified in any way in clause 2.1, it is subject to two qualifications:

(1) The provisions of clauses 6.8 and 6.9 entitling the contractor to terminate its employment under the contract, e.g. if the employer is in default as stated in clause 6.8 or becomes insolvent and subject to the procedural requirements of those clauses.
(2) If the employer – or the architect – prevents the contractor from completing its work by the completion date unless, of course, a proper extension of time has been validly made.

The contractor must carry out its work *in compliance with* the contract documents as defined. The architect must take care to ensure that the description of the work is adequate and this means using precise language. All too often important parts of the contract documents are written in slovenly English. It is impossible to enforce generalisations. The contract documents must contain all the requirements which the employer wishes to impose, and the use of phrases such as 'of good quality' or 'of durable standard' should be avoided. Moreover, the contract documents cannot be used to override or modify the printed conditions because of the last part of clause 1.2, the effectiveness of which has been upheld by the courts on many occasions.

The contractor must complete *all* the work shown in, described by or referred to in the contract documents. The contractor's obligation only comes to an end when the architect has issued the practical completion certificate. Thereafter, the contractor must fulfil its obligations under the defects clause (see section 9.3.1).

The proviso in clause 2.1.2 states that if approval of workmanship or materials is a matter for the architect's opinion, then quality and standards must be to the architect's reasonable satisfaction. The effect of this is discussed in Chapter 4, section 4.2.

There is no further amplification of the contractor's basic obligation although the contractor is, of course, bound to complete the Works *by* the date stated in the contract particulars and if it does not do so, subject to the operation of the provisions for extension of time, it must pay liquidated damages to the employer. In *Glenlion Construction Ltd* v. *The Guinness Trust* (1987), it was held, on slightly different wording ('on or before the date for completion'), that the contractor was entitled to complete early, albeit the architect was not obliged to provide information at times to suit early completion. It is suggested that the contractor is similarly entitled by virtue of the wording 'shall be completed *by*' (emphasis added) under this contract.

It is notable that, unlike the position under MW 98, there is no express requirement for the contractor to proceed with due diligence. This is surprising, particularly in view of the termination provisions in clause 6.4.1.2, which permit the employer to terminate the contractor's employment if it fails to proceed regularly and diligently. This is the same inconsistency which troubled the court in *Greater London Council* v. *Cleveland Bridge & Engineering Co Ltd* (1986) and whose judgment was wholeheartedly endorsed by the Court of Appeal in 1986. There, the court took the view that, although a failure to proceed with due diligence allowed the employer to discharge the contractor, it did not amount to a breach of contract. Whether a term that the contractor must proceed regularly and diligently would be implied in this instance is open to question. However, the employer's ability to terminate on this ground emphasises the importance of understanding its meaning. The phrase 'regularly and diligently' has been defined in *West Faulkner* v. *London Borough of Newham* (1994), by the Court of Appeal:

> 'Although the contractor must proceed both regularly and diligently with the Works, and although each word imports into that obligation certain discrete concepts which would not otherwise inform it, there is a measure of overlap between them and it is thus unhelpful to seek to define two quite separate and distinct obligations.
>
> What particularly is supplied by the word "regularly" is not least a requirement to attend for work on a regular daily basis with sufficient in the way of men, materials and plant to have the physical capacity to progress the Works substantially in accordance with the contractual obligations.
>
> What in particular the word "diligently" contributes to the concept is the need to apply that physical capacity industriously and efficiently towards the same end.
>
> Taken together the obligation upon the contractor is essentially to proceed continuously, industriously and efficiently with appropriate physical resources so as to progress the Works steadily towards completion substantially in accordance with the contractual requirements as to time, sequence and quality of work.'

Whether a contractor is or is not proceeding regularly and diligently clearly depends on all the circumstances, but it is something which most architects instinctively recognise.

The sensible architect will require a contractor to provide a programme, preferably showing logic links, but mere failure to comply with a programme is not usually sufficient alone to show failure to proceed regularly and diligently. The architect will not approve the contractor's programme, because it is essentially a setting out of the contractor's intentions. In any event, it is not thought that approval by the architect has any great significance in this context. It certainly does not make the architect responsible for the correctness of the programme: *Hampshire County Council* v. *Stanley Hugh Leach Ltd* (1991).

5.1.3 Workmanship and materials

Clause 2.1 (clause 2.2.1 in MWD) deals with workmanship and materials. It requires the contractor to use materials and workmanship as specified in the contract documents or, if so stated, in accordance with the architect's satisfaction. MWD expressly deals with the situation where materials and workmanship are not specified nor made subject to the architect's opinion or satisfaction. In such instances, the standard of materials and workmanship are to be appropriate to the Works. If the CDP is concerned, the standard is to be appropriate to the CDP. It is not clear why MW omits this provision in regard to the Works.

This provision in fact imposes an obligation on the architect to define the quality and standards of workmanship and materials very carefully indeed. The employer normally places reliance on the architect, not the contractor, to specify correctly and the contractor is unlikely to be liable if it supplies the wrong material merely because the architect has wrongly indicated that any one of a range of materials may be used: *Rotherham Metropolitan Borough Council* v. *Frank Haslam Milan & Co Ltd and Ano* (1996). Where it is impossible adequately to specify in sufficient detail, the architect could do worse than fall back on the phraseology of the common law:

- All work shall be carried out in a good and workmanlike manner and with proper care and skill.
- All goods and materials shall be of good quality and reasonably fit for their intended purpose.

Quite clearly, the contractor is expected to show a reasonable degree of competence and to employ skilled tradesmen. Although the architect has no power to direct the contractor as to how it should carry out its work, save to the extent that the architect can by a variation order issued under clause 3.6 change the order or period in which the Works are to be carried out, the architect does have power under clause 3.8 to instruct that any unsatisfactory employee or anyone else employed on the Works should be excluded.

Because of the wording of clause 2.1 the contractor *must* provide workmanship, materials and goods of the standards and quality specified. It is no excuse that they are not in fact available, since the contractor does not enjoy the benefit of the limitation of SBC that it need do so only so far as the goods, etc. are procurable. This is a matter which is at the contractor's risk and failure to supply is a breach of contract. Having said that, there is an argument that, if a specified item becomes unavailable, the contract might be frustrated.

Materials and goods are also referred to in clause 4.3 in the context of progress payments. The architect is *bound* to include in progress certificates the value of any materials and goods which have been properly brought to the site and reasonably for the purposes of carrying out the Works, provided that they are adequately protected from the weather or other causes of damage. Use of the words 'properly' and 'reasonably' is presumably to prevent the contractor bringing materials onto site far in advance of the date they are required simply to secure valuation and payment. The architect would be entitled to ignore prematurely delivered materials in any certificate. Indeed, the architect would have no power under the contract to certify the value of such materials.

This is a very dangerous provision from the employer's point of view since the unfixed materials and goods will not necessarily become the property of the employer, even though the employer has paid for them, if, in fact, they are not the employer's property in law, as will often be the case. Builders' merchants frequently include a 'retention of title' clause in their contracts of sale: *Dawber Williamson Roofing Co Ltd* v. *Humberside County Council* (1979). Ideally, clause 4.3 should be amended so as to ensure that the inclusion of the value of unfixed materials is a matter for the architect's discretion. Less satisfactorily, it may include an appropriate provision requiring the contractor's application for progress payments to be accompanied by documentary proof of ownership.

It is probable that the architect is safe from an action for negligence for operating clause 4.3 as it stands because the employer has signed MW which says that the architect *shall* include on-site unfixed materials. However, architects should be alert to the dangers, and the Joint Contracts Tribunal should issue an amended clause 4.3 as a matter of urgency. This plea was first made in the first edition of this book, in 1990 – clearly to no avail.

5.1.4 Statutory obligations

MW clause 2.1.1 (clause 2.1 in MWD) imposes on the contractor a duty to comply with all statutory obligations, e.g. those imposed by the Building Regulations 2001 and subsequent amendments and clause 2.6 (clause 2.7 of MWD) requires payment of all fees and charges which are legally 'demandable'. Clause 2.5 (clause 2.6 in MWD) imposes on the contractor an obligation to give immediate written notice to the architect if the contractor discovers any divergence between the statutory requirements and the contract documents or one of the architect's instructions.

There then follows a provision which expressly excludes the contractor from any liability under the contract if the Works do not comply with statutory requirements. There is an important proviso that the contractor must not have been in breach of its obligations to immediately notify the architect if any divergence was found and that the non-compliance must have resulted from carrying out the Works in accordance with the contract documents or in accordance with an architect's instruction.

The effect of this provision is contractually to exempt the contractor from liability to the employer if the Works do not comply with statutory requirements provided the contractor has carried out the work in accordance with the contract documents or any architect's instruction, where, for example, the contractor does not spot the divergence. The contractor's obligation to notify the architect of a divergence only arises *if* the contractor spots it, and unless the contractor does so is under no obligation to notify the architect. Plainly, it is ineffective to exonerate the contractor from the duty to comply with statutory requirements and if, for example, the contractor carried out work in breach of the Building Regulations, it would be criminally liable since the primary liability to comply with them rests on the contractor: *Perry* v. *Tendring District Council* (1985). The wording is probably sufficiently wide to protect the contractor from any action by the employer. But it does not protect the architect, and if the fault is the architect's the employer will be able to recover any losses from the negligent architect.

5.1.5 Contractor's representative

Clause 3.2 obliges the contractor to keep on the Works a 'competent person in charge' at all reasonable times, i.e. during normal working hours. This person is intended to be the contractor's full-time representative on site, but the appointment and replacement are not subject to the architect's approval although in an appropriate case the architect could exercise powers under clause 3.8 to require the exclusion of the person in charge from the site and thus force the contractor's hand.

Competent means that the contractor's representative must have sufficient skill and knowledge, and it is essential that the architect knows from the outset of this person's identity since he or she is the contractor's agent for the purpose of accepting instructions. Instructions given to the person in charge are *deemed* to be given to the contractor. From the contractor's point of view, such a person must be completely reliable.

Many architects include in the tender documents a requirement that the contractor give adequate notice of the replacement of the person in charge, and this is good contract practice.

5.2 *Other obligations*

5.2.1 Access to the works and premises

There is no express provision in MW or MWD giving the architect and representatives access to the Works at all reasonable times. Such a right of access to the site is implied under the general law. However, there is a gap, because the architect may need access to the contractor's workshops, etc. where items are being prepared for the contract. If the architect does need to visit the contractor's premises to keep an eye on things, there must be an appropriate clause in the contract documents, following the wording of SBC clause 3.1, because it is probable that the architect does not have that right of access to either the contractor's or any subcontractor's workshops under the general law.

5.2.2 Compliance with architect's instructions

Clause 3.4 obliges the contractor to carry out the architect's instructions forthwith and this obligation is not conditioned in any way. Except under MWD, the contractor is given no right to object to architect's instructions, even those involving a variation. Under MWD clause 3.4.2, the contractor's consent is required to any instruction affecting the design of CDP work.

The sanction for non-compliance by the contractor is set out in clause 3.5. If a contractor fails to comply with an architect's instruction, the architect may serve the contractor with a written notice requiring compliance. If the contractor has not complied within seven days from receipt of that notice, the employer may engage others to carry out the work and an appropriate deduction is to be made from the contract sum. Figure 4.9 is a suggested letter from the architect, and Figure 4.10 is a possible reply.

To avoid the possibility of an argument that the employer has waived its rights, it is essential that the architect ensures that the machinery provided is put into operation if the contractor fails to comply with instructions.

MW and MWD contain no express provision for the architect to issue instructions for the removal of work not in accordance with the contract as does SBC clause 3.18.1, but it is probable that the broad wording of clause 3.4 extends to cover that situation.

5.2.3 Suspension of obligations

Under the provisions of clause 4.7 the contractor has the right to suspend performance of its obligations under the contract under certain circumstances. In effect, this means that the contractor may cease work on site and elsewhere. In order to be able to do so, certain criteria must be satisfied. First, the employer must have failed to pay money due by the final date for payment. Second, the contractor must have given seven days' notice in writing of its intention. The contractor must resume work as soon as it is paid in full. This provision puts the relevant provision in the Housing Grants, Construction and Regeneration Act 1996 into effect. There is no ordinary common law right to suspend work for late payment although, in practice, it is often done. A contractor who suspends work legitimately will be entitled to an extension of time, but not, it seems, loss and expense. Claiming damages for breach of contract appears to be the only, but by no means easy, option.

5.2.4 Other rights and obligations

Tables 5.1 and 5.2 summarise the contractor's rights and duties generally. Other matters referred to in those tables are dealt with in the appropriate chapters.

5.3 *Summary*

Contractor's obligations

MW and MWD impose certain express obligations on the contractor; other obligations are implied at common law:

- The contractor must carry out and complete the Works in accordance with the contract documents, the Health and Safety Plan and statutory requirements
- The contractor must do this by the specified completion date
- The contract documents must be precise and specify quality and standards of both workmanship and materials
- The contractor must comply with statutory obligations and pay all fees and charges involved
- *If* the contractor discovers a divergence between the statutory requirements and the contract documents or any architect's instruction it must give the architect written notice immediately
- Subject to this the contractor is exempted from liability to the employer if the Works as built contravene statute law
- The contractor must keep a competent representative on site during normal working hours and any instructions given by the architect to that representative are deemed to have been given to the contractor
- The contractor must comply with all architect's instructions issued under the contract
- The employer has an option to engage others to carry out architect's instructions should the contractor not carry them out within seven days of a written notice from the architect.
- The contractor may suspend work on seven days' notice if the employer fails to pay.

CHAPTER SIX
EMPLOYER

6.1 *Powers and duties: express and implied*

Like those of the contractor, some of the employer's powers and duties arise from the express provisions of MW and MWD, while others arise under the general law.

Tables 6.1 and 6.2 set out those powers and duties of the employer which arise from the express terms of the contract.

This section is concerned with the obligations which are placed on the employer by way of implied terms. These are provisions which the law writes into a contract in order to make it commercially effective. Terms will be implied to the extent that they are not inconsistent with the express terms which may also exclude or modify the implied terms.

It is an implied term of MW and MWD that the employer will do all that is reasonably necessary to bring about completion of the contract: *Luxor (Eastbourne) Ltd* v. *Cooper* (1941). Conversely, it is implied that the employer will not so act as to prevent the contractor from completing in the time and in the manner envisaged by the agreement: *Cory* v. *City of London Corporation* (1951). Breach of either of these implied terms which results in loss to the contractor will give rise to a claim for damages at common law and the contractor can pursue that claim under the arbitration agreement.

Equally, if the employer – either personally or through the agency of the architect or that of anyone else for whom the employer is responsible in law – hinders or prevents the contractor from completion in due time, not only is the employer in breach of contract, but he or she will be disentitled from enforcing the liquidated damages clause.

The various cases put the duty in different ways, but in essence the position can be summarised as follows:

- The employer and the employer's agents must do all things necessary to enable the contractor to carry out and complete the Works expeditiously and in accordance with the contract.

Table 6.1
Employer's powers under MW and MWD

Clause	Power	Comment
2.8.1 (2.9.1 under MWD)	Deduct liquidated damages from any monies due to the contractor under this contract or recover them as a debt	If the Works are not completed by the completion date or any extended date and provided that a notice of deduction has been given
3.1	Assign the contract	If the contractor consents
3.5	Employ and pay others to carry out the work An appropriate deduction is to be made from the contract sum	If the contractor fails to comply with a written notice from the architect requiring compliance with an instruction and within seven days from its receipt
4.6.2	Give written notice to the contractor specifying amount to be withheld or deducted	The notice must state the amount, the grounds and the amount in respect of each ground and must be given not less than five days before the final date for payment of a progress payment
4.8.3	Give written notice to the contractor specifying amount to be withheld or deducted	The notice must state the amount, the grounds and the amount in respect of each ground and must be given not later than five days before the final date for payment of the amount in the final certificate
5.5	Reasonably require the contractor to produce evidence of insurance referred to in clauses 5.3 and 5.4A if applicable	
6.4.2	By written notice terminate the contractor's employment under the contract	If the contractor has not ended the specified default within seven days of receiving a default notice
6.5.1	By written notice terminate the contractor's employment under the contract	At any time if the contractor is insolvent
6.5.2.3	Take reasonable measures to ensure that the site, the Works and materials on site are protected and not removed	The contractor must allow and not hinder the measures

Table 6.1 *Contd*

Clause	Power	Comment
6.6	By written notice terminate the contractor's employment under this or any other contract	If the contractor is guilty of corruption
6.7.1	Engage others to carry out and complete the Works With others, take possession of the site Use all temporary buildings, etc.	After the employer has terminated under clauses 6.4 to 6.6
6.10.1	Give written notice warning that if suspension is not ended within seven days, termination may follow	The notice must be served on the contractor by special, recorded or actual delivery Notice may be served if the whole or substantially the whole of the Works is suspended for one month due to: • *force majeure; or* • architect's instructions as a result of negligence or default of statutory authority; *or* • loss by specified perils; *or* • civil commotion or terrorism; *or* • exercise of UK government power
	Terminate the contractor's employment under the contract	By notice served on the contractor by special, recorded or actual delivery if the suspension does not end
7.1	Agree to resolve the dispute by mediation	
Schedule 1 para 1	By written notice jointly with the contractor to the arbitrator stating that they wish the arbitration to be conducted in accordance with any amendments to the JCT 2005 CIMAR	
Schedule 1 para 2.1	Serve on the contractor a written notice	If the employer wishes a dispute to be resolved by arbitration
Schedule 1 para 2.3	Give a further arbitration notice to the contractor referring to any other dispute	After the arbitrator has been appointed Rule 3.3 applies

Table 6.2
Employer's duties under MW and MWD

Clause	Duty	Comment
2.8.3 (2.9.3 under MWD)	Inform the contractor in writing if there is an intention to deduct liquidated damages from the final certificate sum	Not later than the issue of the final certificate
3.9.1	Ensure that the planning supervisor carries out his duties Ensure that the principal contractor carries out its duties	 If the contractor is not the principal contractor
3.10	Immediately notify the contractor in writing of the name and address of the new appointee	If the employer replaces the planning supervisor or the principal contractor
4.1	Pay to the contractor any VAT properly chargeable	
4.3	Pay to the contractor amounts certified by the architect	Payment must be made within 14 days of the date of the certificate issued under clause 4.3
4.4	Pay simple interest at 5% above Bank of England base rate to contractor on amount not properly paid, until paid in full	If employer fails to pay any amount due by final date for payment
4.5	Pay to the contractor amounts certified by the architect	If the architect issues a certificate under clause 4.5 within 14 days of the date of practical completion
4.6.1	Give written notice to contractor specifying amount to be paid in respect of amount stated as due in any certificate	Not later than five days after issue of certificate of payment
4.6.3	Pay the amount stated as due in any certificate or notice under clause 4.6.1 (if issued)	If the employer does not give notice under clause 4.6.2

Table 6.2 *Contd*

Clause	Duty	Comment
4.8.2	Give written notice to the contractor specifying amount to be paid in respect of amount stated as due in final certificate	Not later than five days after issue of final certificate
4.8.4	Pay the amount stated as due in the final certificate or notice under clause 4.8.2 (if issued)	If the employer does not give notice under clause 4.8.3
4.9	Pay simple interest at 5% above Bank of England base rate to the contractor on amount not properly paid in the final certificate, until paid	If the employer fails to pay any amount he is due to pay by the final date for payment
5.4A.2.3	Pay insurance monies to the contractor	On certification by the architect
5.4B.1	Maintain adequate insurances against the specified perils	Existing structures, contents and the Works
5.5	Produce evidence of such insurances	If the contractor so requires
6.7.3	Set out an account in a statement	Within a reasonable time after completion of the Works and making good of defects if the employer has terminated and the architect has not set out an account in a certificate
6.11.4	Pay to the contractor the full amount properly due in respect of the contractor's account	If the contractor's employment is terminated under clauses 6.8 to 6.10 Payment must be made within 28 days of submission of the account by the contractor

- Neither the employer nor the employer's agents will in any way hinder or prevent the contractor from carrying out and completing the Works expeditiously and in accordance with the contract.

The scope of these implied terms is very broad and in recent years more and more claims for breach of them have been before arbitrators and the courts. The employer must not, for example, attempt to give direct orders to the contractor, and must see that the site is available to the contractor on the date specified in clause 2.1 and that access to it is unimpeded by those for whom the employer is responsible or over whom the employer exercises effective control. For example, in *Rapid Building Co Ltd* v. *Ealing Family Housing Association Ltd* (1985), squatters had occupied part of the site and as a result the employers were unable to give possession on the due date. This was a breach of contract which caused appreciable delay to the contractors who were held to be entitled to damages. This is especially important in the case of work to existing structures or occupied buildings, and these are matters which should be discussed between the employer and the professional advisers before the contract is let.

6.2 *Rights under MW and MWD*

6.2.1 General

Although the contract is between the employer and the contractor – who are the only parties to it – an analysis of the contract clauses shows that the employer has few express rights of any substance.

The employer's major right is to have the Works contracted for handed over by the agreed completion date, properly completed in accordance with the contract documents. The employer also has the right to assign the benefits of the contract to a third party, provided the contractor consents in writing (clause 3.1), although it is difficult to envisage many circumstances in which an employer could wish to do that!

6.2.2 Damages for non-completion

If the contractor does not complete the Works by the agreed completion date, the employer is entitled to recover liquidated damages at the rate specified in the contract particulars for each complete week, day or other specified period during which the works remain uncompleted after the original or extended completion date.

There is no requirement that the architect should issue a certificate of non-completion. Although it is good practice for the architect to notify the employer in writing when the completion date has passed, a copy can be sent to the contractor (a standard form is available for this purpose). This reminds both parties that liquidated damages are now payable. The mere fact of late completion is sufficient to bring clause 2.8 (2.9 in MWD) into operation. The employer is now given an express right to deduct liquidated damages from monies due to the contractor, but if (unusually) no sums are due or are to become due to the contractor, the employer must take action to recover them as a debt.

Before any money can be deducted from money due to the contractor, the employer must have given written notice under clauses 4.6.2 or 4.8.3 to comply with the Housing Grants, Construction and Regeneration Act 1996. Inexplicably, the contract states that the employer must also inform the contractor in writing not later than the date of the issue of the final certificate. This duty is stated to be additional. Therefore, if the employer, unusually, waits until the final certificate to deduct liquidated damages, one notice must be served not later than five days before payment is finally to be made and an earlier notice when the final certificate is issued. Were it not for the use of the word 'additionally', a single notice on the issue of the final certificate would have been enough to satisfy the provision.

In practice, the architect should, of course, advise the employer of all rights before the contract overruns. Waiting until the final certificate to deduct liquidated damages may not be a good idea.

6.2.3 Other rights

These are summarised in Table 6.1 and are there described as 'powers'. They are discussed in the appropriate chapters.

6.3 Duties under MW and MWD

6.3.1 General

The essence of a duty is that it is something that must be done. It is not permissive; it is mandatory. Breach of a duty imposed by the contract will render the employer liable to an action for damages by the contractor in respect of any proven loss.

Some breaches of contract may, in fact, entitle the contractor to treat the contract as at an end. They are called 'repudiatory breaches' which means that they go to the basis of the contract. For example, physically expelling the contractor and its operatives from site would be a repudiatory breach because the employer is effectively demonstrating a wish no longer to be bound by the contract. Alternatively, for the contractor to walk off site permanently before the Works are complete would be a repudiatory breach on its part.

Breach of any contractual duty on the part of the employer will always, in theory, entitle the contractor to at least nominal damages, although in some cases any loss will be difficult if not impossible to quantify.

6.3.2 Payment

From the contractor's point of view, the most fundamental duty of the employer is to make payment in accordance with the terms of the contract. However, while steady payment against certificates is essential from the contractor's viewpoint, the general law does not usually regard non- or late payment as such as a major breach of contract. Certainly non-payment of one certified sum to the contractor is not generally regarded as repudiation, but in *D. R. Bradley (Cable Jointing) Ltd* v. *Jefco Mechanical Services Ltd* (1988), a subsequent refusal to pay on a certificate did go to the root of the contract since 'it reasonably shattered [the contractor's] confidence in being paid'. *C. J. Elvin Building Services Ltd* v. *Peter & Alexa Noble* (2003) is a more recent case which supports the view that repeated failures to pay amount to repudiation. Case law also suggests that a certificate is not as good as cash, as most contractors appear to think, since in some limited cases the employer may be justified in withholding payment of certificated amounts pending adjudication or arbitration: *C. M. Pillings & Co Ltd* v. *Kent Investments Ltd* (1986); *R. M. Douglas Construction Ltd* v. *Bass Leisure Ltd* (1991). That is particularly so since the introduction of specific notices to be given if the employer wishes to withhold payment or deduct any amounts in respect of an interim progress payment (clause 4.6.2) or the final payment (clause 4.8.3) following the Housing Grants, Construction and Regeneration Act 1996. The employer must have a substantial reason for withholding payment and the grounds must be indicated and the amount proposed to be withheld in respect of each ground, otherwise the contractor may be justified in suspending work under clause 4.7 or giving a default notice prior to termination under clause 6.8.1. Clauses 4.4 and 4.9 now contain provisions requiring the employer to pay simple interest on amounts not properly paid.

MW and MWD are quite specific in their terms as to payment (see Chapter 11). The basic provisions are to be found in clauses 4.3 to 4.9 inclusive.

It is essential that the employer is made aware of the need to pay promptly in accordance with the contract terms because nothing sours good working relationships more than late payment, and most contractors do have a cash-flow problem. In these circumstances the architect ought to write to the employer pointing out the contractual position (Figure 6.1).

Once the architect has issued a payment certificate under clauses 4.3 and 4.5 (progress payments and penultimate certificate) the employer is given a period in which to honour the certificate. Payment must be made *within 14 days of the date of the certificate,* which means payment before the expiry of that period. This is a very tight timetable and the architect should send the certificate to the employer on the day it is issued. Assuming no postal delays (which cannot be relied on these days) and dispatch by first-class post, the employer effectively has no more than 13 days for paying what is due. Delivery by hand or by 'next day' special delivery is to be preferred. The employer must give notice in writing (clause 4.6.1) to the contractor within five days from the date of issue of a certificate. The notice must state the amount of payment to be made in respect of the amount stated as due in the certificate, to what it relates and the basis of calculation. This appears to give the employer the chance to abate the certified sum. The second notice the employer may give relates to any amount withheld from the amount due. However, it should be noted that the contractor's entitlement to suspend work if the employer does not pay the amount certified is subject only to the second notice (clause 4.6.2). The termination provision in clause 6.8.1.1 is less precise, simply referring to a failure to 'pay by the final date for payment the amount properly due to the Contractor in respect of any certificate...', leaving open the question of whether the amount properly due is the same as the certified sum.

Clause 4.6.3, however, is a rewording of the equivalent clause in MW 98 and states quite clearly that if no withholding notice is given the employer must pay the amount stated in the clause 4.6.1 notice. Only if this notice is not given must the employer pay the amount certified. This seems to place beyond doubt that the employer can abate the certified amount although the precise details of a clause 4.6.1 notice would be subject to review by an adjudicator or arbitrator.

The same timetable effectively applies to the final certificate under clause 4.8, though the wording in that case is to the effect that the final date for payment of the amount by employer to the contractor or by the contractor to the employer is to be 14 days from the date of issue of the certificate.

Figure 6.1
Letter from architect to employer if employer is slow to honour certificates

Dear Sir

PROJECT TITLE

The contractor has complained to me of not receiving payment due on certificates until after the period allowed in the contract.

Under the terms of the contract, you have fourteen days from the date of the certificate within which to make payment. Failure to pay within the stipulated time entitles the contractor to suspend performance of its obligations under the contract or terminate its employment under the contract. If the contractor terminated its employment, the consequences would be extremely expensive. Sums outstanding attract simple interest at the rate of 5% above Bank of England Base Rate.

Quite apart from the strict legal requirements as to payment, it is good practice to pay promptly because the contractor is always in the position of having paid out substantial sums well before payment is due. Prompt payment is crucial to its cash flow and, consequently, late payment spoils good working relations.

If you would bear this in mind, there should be benefits on all sides.

Yours faithfully

Payment by cheque is probably good payment, but it is no excuse for the employer to say, for example, that the computer arrangements do not fit in with the scheme of certificates, which is an increasingly common excuse for late payment. If this is indeed the case, the payment period, or better still the computer arrangements, should have been amended before the contract was let.

The employer is entitled to retain retention on progress payments and, in the case of a contract overrun, can set off amounts due as liquidated damages under clause 2.8 (clause 2.9 in MWD) provided the complex notice provisions are satisfied.

6.3.3 Retention

The employer has certain rights in the retention percentage, which is commonly 5%. The percentage of work to be valued and certified up to practical completion and within 14 days afterwards is to be inserted in the contract particulars. The default entries are 95% and $97\frac{1}{2}$% respectively. It is not stated that the retention is trust money and it is likely, therefore, that the contractor is at risk if the employer becomes insolvent.

The employer's rights in the retention are to have it as a fund from which defects may be remedied and other bona fide claims settled. But morally the retention is the contractor's money and once again it is suggested that it would be an improvement if the JCT would amend clause 4.3 to provide that the employer's interest in the retention was 'fiduciary as trustee for the contractor'.

6.3.4 Other duties

These are summarised in Table 6.2 and are commented on as appropriate in other chapters.

6.4 *Summary*

Rights and duties

- The employer must not hinder or prevent the contractor from carrying out and completing the Works as envisaged by the contract. Breach of this duty will render the employer liable to an action for damages and may disentitle the employer from enforcing the liquidated damages clause.

- Should the contractor fail to complete the Works by the completion date, the employer is entitled to recover liquidated damages.
- The employer must pay the contractor on the architect's certificates, payment being due within 14 days of the date of each certificate.
- The retention fund may be used by the employer to satisfy bona fide and quantified claims but it is the contractor's money and not the employer's.

CHAPTER SEVEN
QUANTITY SURVEYOR AND CLERK OF WORKS

7.1 *Quantity surveyor*

7.1.1 Appointment

MW and MWD make no provision for the appointment of a quantity surveyor. Nevertheless, the architect may wish to have the assistance of a quantity surveyor in valuing the Works.

The decision whether or not to appoint a quantity surveyor will be taken by the employer with the architect's advice. The advice will naturally take into account the amount of work which the quantity surveyor could be asked to carry out. In the case of small works, the use of a quantity surveyor tends to be the exception. The architect should be competent to deal with interim payments, the computation of the final sum and valuation of variations if the work is of a simple and straightforward nature, particularly if a clause is inserted to provide for stage payments. If, however, the architect has any doubts about competence in this field or the work is complex, the use of a quantity surveyor is indicated. It is important to remember that, if architects hold themselves out to their clients as capable of carrying out quantity surveying functions, that is the standard which will be expected of them. The architect should put forward the architect's name as quantity surveyor unless the position has been checked with the professional indemnity insurers. Since there is no mention of the quantity surveyor in the contract, there appears to be no valid reason why the architect's name should ever be put forward. There is no benefit in so doing, and there may be insurance problems.

If the architect decides that a quantity surveyor is required for the works, the employer should be informed at the earliest possible opportunity, usually stage A – Appraisal, so that the quantity surveyor can assist throughout the project. The architect should take care to explain to the employer why a quantity surveyor is required (Figure 7.1).

Figure 7.1
Letter from architect to employer if the services of a quantity surveyor are required

Dear Sir

PROJECT TITLE

In view of the particular nature/size/value [*delete as appropriate*] of this project, the services of a quantity surveyor will be required. A quantity surveyor is the specialist in building economics and will deal with preparation of valuations for progress payments and variations and provide overall cost advice. It is important that a quantity surveyor should be appointed at this stage so that you can derive the maximum benefit from the advice and the quantity surveyor can become fully involved with the project.

Although I cannot recommend any particular quantity surveyor, you may wish to consider the use of [*insert name*] of [*insert address*], with whom I have worked many times in the past. If you will let me have your agreement, I will carry out some preliminary negotiations on your behalf and advise you regarding the letter of appointment.

The use of consultants is covered by clauses 1.6.4, 2.6, 3.8, 3.11 and 7.3 inclusive of the Conditions of Engagement CE/99, a copy of which is already in your possession.

Yours faithfully

The situation will be simplified if the employer has been given a copy of the RIBA Conditions of Engagement SW/99 or CE/99 on which, presumably, the architect's engagement is based. It is worth noting at this point that the architect occasionally may be engaged under the Standard Form of Agreement for the Engagement of an Architect (SFA/99), which is a somewhat more complex set of terms. Where the client is a consumer, a common situation where the main contract is MW or MWD, care must be taken to explain and individually negotiate each term whichever standard conditions of engagement are used. Otherwise, the terms may be invalid under the Unfair Contract Terms in Consumer Contracts Regulations 1999: *Picardi* v. *Mr & Mrs Cuniberti* (2003).

7.1.2 Duties

Because the contract does not provide for the appointment of a quantity surveyor, it is no surprise that there are no duties allocated to the quantity surveyor. If it is decided to make such an appointment, it is sensible to include a clause in the contract to that effect. Although terms may be implied to cover the associated duties, it is preferable to deal with powers and duties of the quantity surveyor by means of a specially worded and inserted provision. There are two possible ways of dealing with this:

(1) The insertion of a brief clause in the contract to cover the activities the architect intends the quantity surveyor to perform
(2) A letter to the contractor informing him that the quantity surveyor is to be the architect's authorised representative in respect of specified activities.

The introduction of a clause in the contract is probably the more satisfactory way of accomplishing the objective. The contractor then knows, at tender stage, just what is intended. If the architect handles the matter by means of a letter at the beginning of the contract, the contractor might object violently to it and relations will have been soured at the start. Another danger is that the contractor may not appreciate the limits of the quantity surveyor's duties and may carry out as instructions what are simply the quantity surveyor's comments on some aspect of the work. Although the contractor would be wrong to do so, it is little consolation to the employer if disruption and delay results. If it is felt that the quantity surveyor's duties must be dealt with in this way, great care must be taken with the letter (Figure 7.2).

Figure 7.2
Letter from architect to contractor regarding the duties of the quantity surveyor

Dear Sir

PROJECT TITLE

The quantity surveyor named in additional clause [*insert number*] of the contract is [*insert name*] of [*insert address*]. This letter is to notify you that the quantity surveyor will be my authorised representative to carry out the duties of valuation, calculation, checking, computation and measurement in respect of the following clauses of the Conditions of Contract:

2.4 [2.5 when using MWD]
2.10 [2.11 when using MWD]
3.5
3.6
3.7
4.3
4.5
4.8
6.7.3
6.11

You must supply all necessary documents, vouchers, etc. to enable the quantity surveyor to carry out this work. Your attention is drawn to the fact that the quantity surveyor's duties are limited to quantification. The quantity surveyor is not empowered to issue instructions, certificates, make awards or decide liability for payment.

Yours faithfully

Copies: Employer
Quantity surveyor

If a clause is included, proper legal assistance must be obtained for its drafting. Among the duties that the architect wishes the quantity surveyor to carry out may be the following:

- Value the variation in the unlikely event that an inconsistency results in a variation under clause 2.4 (clause 2.5 in MWD)
- Value the amount of the employer's contribution under clause 2.10 (clause 2.11 in MWD), if the architect instructs otherwise than that the contractor should make good all defects at its own expense (see section 9.3.4)
- Valuation work under clause 3.5 if the employer has to employ others to carry out work contained in architect's instructions
- Valuation of variations in accordance with clause 3.6
- Valuation of instructions in connection with provisional sums in clause 3.7
- Measurement and valuations under clause 4.3
- Measurement and valuations under clause 4.5
- Computation of the final sum under clause 4.8
- Valuations under clause 6.7.3
- Checking the contractor's account prepared under clause 6.11.

Provision should be included for the quantity surveyor to require the contractor to supply any necessary documents for the purpose of carrying out the valuations, calculations or measurements. It is also prudent to stress that the quantity surveyor's duties cover quantification only with no power to agree or decide liability for payment on or to issue instructions, certification or make any awards and this would be the situation under the general law: *John Laing Developments Ltd* v. *County and District Properties Ltd* (1982).

7.1.3 Responsibilities

The quantity surveyor's responsibility is to the employer in contract and in tort: *Wessex Regional Health Authority* v. *HLM Design and Webb* (1994). Even if a list of duties is included in a special clause inserted in MW and MWD, the quantity surveyor will not be liable to the contractor in contract because, like the architect, the quantity surveyor is not a party to the contract. There is a remote possibility of liability to the contractor in tort but the architect is the professional entrusted with the task of certifying all payments. If the quantity surveyor makes a mistake, the architect must correct it. The onus on the architect to be satisfied about all the quantity surveyor's calculations so far as it is reasonable to do so cannot be emphasised too much.

Clause 3.11.1 of the RIBA Conditions of Engagement CE/99 states that, where the client employs a consultant, that consultant will be held responsible for the competence, performance of services and visits to the site in connection with the work undertaken by that consultant. The forerunner of this clause in the RIBA Conditions of Engagement was upheld in *Investors in Industry Commercial Properties Ltd* v. *South Bedfordshire District Council* (1986) where the Court of Appeal said that following the appointment of a consultant by the client on the architect's recommendation:

> 'the architect will normally carry no legal responsiblity for the work done by the expert which is beyond the capability of an architect of ordinary competence... [but] this is subject to one important qualification. If any danger or problem arises in connection with the work allotted to the expert of which an architect of ordinary competence reasonably ought to have been aware and reasonably could be expected to warn the client despite the employment of the expert, and despite what the expert says or does about it, it is... the duty of the architect to warn the client about it.'

However, if the quantity surveyor makes a mistake, the architect will always be the first target for the client's displeasure.

The quantity surveyor will be liable for any advice given directly to the employer if the advice is negligently given. In practice, this will happen only rarely. An example might be if the quantity surveyor is present at the meeting to open tenders for the main contract and volunteers negligent advice, as a direct result of which the employer enters into a contract with a contractor which turns out to be more expensive than the employer was led to believe. Even in this sort of instance, there is a very good chance that the employer will blame the architect for the difficulty. The employer may have a point, but everything will turn on the precise circumstances in which the quantity surveyor gave the advice. The quantity surveyor has exactly the same sort of contractual relationship with the employer as the architect has. The architect's duties are different and include, in addition, the authority and duty to co-ordinate and integrate the consultant's services and generally to co-operate as reasonably necessary for the carrying out of the services (Conditions of Engagement CE/99, clause 2.6), but the architect can never be responsible for the quantity surveyor's actions or defaults.

Architects should resist, as far as possible, any efforts on the part of the employer to get them to appoint quantity surveyors themselves. From the employer's point of view it is understandable that it is preferable to deal with everything through one person, but the arrangement leaves the architect very exposed. The modern practice in negligence cases seems to be to sue everyone in sight and some commentators argue that it makes no real difference whether the employer appoints all consultants directly or simply appoints the architect and leaves the architect to appoint any other consultants, with the employer's permission, who may be necessary. That is thought to be a wrong view. Figure 7.3 indicates the various relationships which are possible. If the employer suggests that the architect appoint the quantity surveyor, it is suggested that the position should be placed on record (Figure 7.4).

It should be remembered that the architect has no contractual relationship with the quantity surveyor (if the employer has appointed one); if there is any liability between architect and quantity surveyor it will be in tort.

7.2 Clerk of works

7.2.1 Appointment

MW and MWD make no provision for a clerk of works. This is because, quite clearly, on most Works for which MW and MWD are suitable, the employment of a clerk of works in a full or part-time capacity is hardly justified. Every project has its own difficulties, however, and if it is considered that a clerk of works is required, the architect must advise the employer accordingly.

The Conditions of Engagement CE/99 expressly state in schedule 2 that the architect will make visits to the construction Works. If it becomes clear that more frequent or even constant inspections are required, the architect may recommend the appointment of a clerk of works (clause 3.10), referred to in CE/99 as 'Site Inspectors'. If the employer is one of those organisations which employ clerks of works on their permanent staffs, that is an excellent arrangement. To avoid misunderstandings, the position should be put on record (Figure 7.5). In the case of most employers, however, it will be a one-off arrangement, the clerk of works being engaged by the employer on a full-time or part-time basis for the duration of the contract. Some firms of architects employ their own clerks of works on a permanent basis, but, in the light of case law, the architect is advised to avoid employing a clerk of works, even

Figure 7.3
Some possible contractual relationships

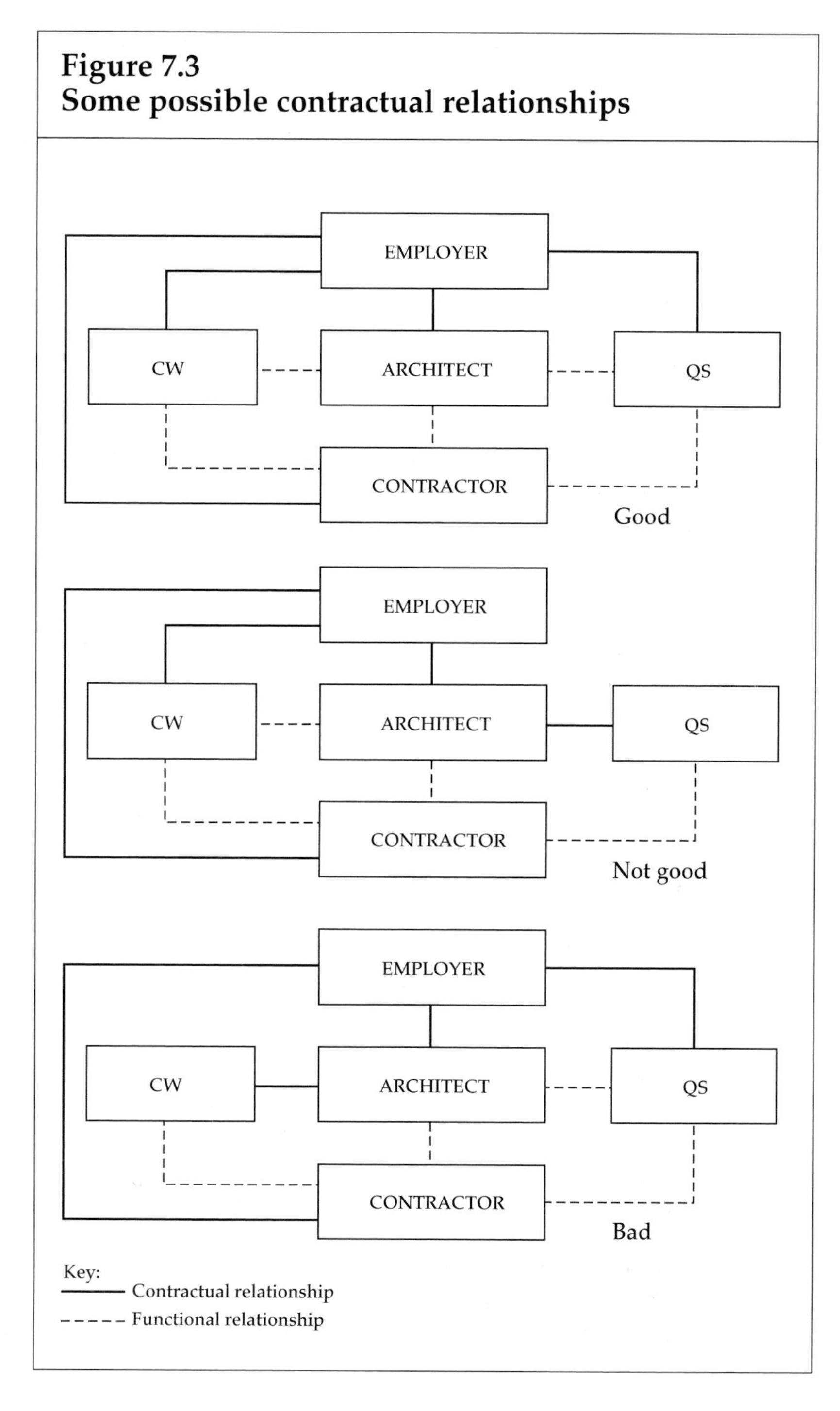

Figure 7.4
Letter from architect to employer if he or she requires the architect to appoint the quantity surveyor

Dear Sir

PROJECT TITLE

Thank you for your letter of the [*insert date*] and I am pleased that you have agreed to the appointment of [*insert name*] as quantity surveyor for this project.

I strongly advise you to appoint the quantity surveyor directly yourself. That is normal practice in construction projects and gives you direct right of access if you should so wish. I cannot provide these services myself and, if you wish me to appoint, I should do so simply as your agent.

I will, in any case, co-ordinate all professional work. The quantity surveyor's fees are the same whether appointed directly, as I advise, or through my office. If you let me know that you will take my advice, I will draft an appropriate letter of appointment for you to send.

Yours faithfully

Figure 7.5
Letter from architect to employer regarding the appointment of a clerk of works

Dear Sir

PROJECT TITLE

In view of the nature/size/value [*delete as appropriate*] of this project, I advise the appointment of a clerk of works on a part-time basis, say [*insert number*] hours per week.

The RIBA Conditions of Engagement CE/99, clause 3.10, provides that where it is agreed that a site inspector is to be appointed, you will appoint and pay that person under a separate agreement. By 'site inspector' CE/99 refers to any kind of site inspector. In this instance a clerk of works is appropriate. A clerk of works will be able to make inspections of such frequency as should ensure the proper carrying out of the work. Such an appointment is likely to repay its cost several times over in savings on lost time and money as the contract progresses.

Although, naturally, I will carry out my own duties with reasonable skill and care, on a contract of this type I cannot accept responsibility for such defects as would be discovered by the employment of a clerk of works.

Yours faithfully

though the employer is prepared to pay an additional sum to cover the clerk of works' salary (see section 7.2.3).

If a clerk of works is to be used, it is essential to include a clause to that effect in the contract. It is suggested that the clause follows the lines of clause 3.3 of the Intermediate Building Contact (IC). That provides that the employer is entitled to appoint a clerk of works, but that the duty of the clerk of works is simply to act as an inspector on the employer's behalf although under the architect's direction.

This states all that is necessary and removes any uncertainty in the contractor's mind regarding the directions of the clerk of works – none are empowered. If the clerk of works attempts to issue instructions, the contractor should write to the architect accordingly (Figure 7.6).

In the interests of the project, the architect should advise the employer to appoint the clerk of works as soon as the contractor's tender has been accepted. This gives the clerk of works time to become thoroughly familiar with the drawings, specification and schedules. The architect will hold a meeting with the clerk of works to brief the latter about the work to be done, but the architect would be wise to confirm the main points to the clerk of works by letter (Figure 7.7), which should outline the aspects of the job considered to be important and act as a reminder of the extent and limitations of the clerk of works' duties.

7.2.2 Duties

If a clause is incorporated based upon IC, clause 3.3, the clerk of works' sole duty is to inspect the Works. The clause makes clear that this function is carried out on behalf of the employer; the clerk of works is not the architect's inspector. This is vitally important as will be seen in the next section. The employer, however, has no power to direct the clerk of works. That is the architect's prerogative alone.

The architect's directions to the clerk of works will presumably embrace how, where and at what intervals inspections should be carried out and to what particular attention should be paid. If the architect issues any directions which are other than purely routine in nature, they should always be confirmed in writing to protect the architect's position. In practice, the clerk of works will often do more than simply inspect. The architect may seek the clerk of works' advice, as a person of some experience, from time to time. The contractor often asks the clerk of works for assistance in solving site problems. Note, however, that a contractor who acts on the advice or even instructions of the clerk of works, does so at its peril. In giving advice, the clerk of works is acting in a personal capacity.

Figure 7.6
Letter from contractor to architect regarding directions from the clerk of works

Dear Sir

PROJECT TITLE

The clerk of works has issued direction number [*insert number*] dated [*insert date*] on site, a copy of which is enclosed.

Such directions have, of course, no contractual effect. Clearly, the directions of the clerk of works issued in relation to the correction of defective work can be very helpful. We are anxious to avoid misunderstandings on site and in this spirit we suggest that the clerk of works should issue no further directions, other than those relating to defective work. There is no reason why all other matters cannot be referred directly to you by telephone and, if appropriate, you can issue an architect's instruction as empowered by the contract.

In our view, the above system would remove a good deal of the uncertainty which must result from the present state of affairs. We look forward to hearing your comments.

Yours faithfully

Figure 7.7
Letter from architect to clerk of works setting out duties

Dear Sir

PROJECT TITLE

My client [*insert name*] has confirmed your appointment as clerk of works for the above contract. I should be pleased if you would call at this office on [*insert date*] at [*insert time*] to be briefed on the project and to collect your copies of drawings, specification, schedules, weekly report forms and site diary.

The contractor is expected to take possession of the site on the [*insert date*]. You will be expected to be present on site for a minimum of [*insert number*] hours per week from that date. I will discuss the timing of your visits when we meet. Let me know at the end of the first week if proper accommodation is not provided for you as described in the specification.

Your duties will be as indicated in the Conditions of Contract special clause [*insert number*], a copy of which is enclosed for your reference. In particular, I wish to draw your attention to the following:

(1) You will be expected to inspect all workmanship and materials to ensure conformity with the contract, i.e. the drawings, specification, schedules and further instructions issued from this office. Any defects must be pointed out to the person in charge, to whom you should address all comments. If any defects are left unremedied for 24 hours or if they are of a major or fundamental nature, you must let me know immediately by telephone.
(2) Although it is common practice for clerks of works to mark defective work on site, you must not make such marks or in any way deface materials on site.
(3) It is not my policy to issue lists of defects to the contractor before practical completion (commonly known as 'snagging lists'). They are open to misinterpretation and should be compiled by the person in charge. Confine yourself to oral comments.
(4) The architect is the only person empowered to issue instructions to the contractor.

Figure 7.7 *Contd*

(5) Any queries, unless of a minor explanatory nature, should be referred to me for a decision. You are not empowered to vary or omit work.
(6) The report sheets must be filled in completely and a copy sent to me on Monday of each week. Pay particular attention to listing all visitors to site and commenting on work done in as much detail as possible.
(7) The diary is provided for you to enter your daily comments.
(8) Remember that your weekly reports and site diary may be called in evidence should a dispute arise, so you must bear this in mind when making your entries which should be as full as possible.

The successful completion of the contract depends in large measure on your relationship with the contractor. If you are in any doubt about anything, please let me know.

Yours faithfully

The architect should make sure that the contractor understands the position at the beginning of the contract. It is prudent to give some time to this topic at the first contract meeting and give some space to it in the minutes. Contractors get into the habit of accepting the clerk of works as speaking on behalf of the architect. Although there is no foundation for this view in the contract, it saves much bad feeling to put the matter beyond doubt. One of the worst things that can happen on a contract is for the architect to have to overrule the clerk of works.

The duties of the clerk of works can, thus, be seen in two parts:

(1) Duties under the contract
(2) Duties by virtue of the architect's specific directions.

The clerk of works' duties under the contract define relations with the contractor and the duty to comply with the architect's directions defines relations with the architect. That is to put the matter in broad terms. Every architect has particular views about the relationship with the clerk of works, but among the duties the clerk of works might be expected to fulfil are the following:

- Inspect the works
- Relay queries and problems back to the architect
- Complete report sheets
- Complete daily diary
- Take measurements as directed
- Take particular notes of such things as portions of the work opened up for inspection under clause 3.4.

The clerk of works is not empowered to put any marks on defective portions of the work. Once such work is removed, it is the property of the contractor who is entitled to expect that the clerk of works will not do any damage to the Works, no matter how slight. If this problem does occur, the contractor should write to the architect immediately (Figure 7.8).

It is, unfortunately, common practice for the clerk of works to issue 'snagging lists' to the contractor, particularly towards the end of a job. There are very rare occasions when there may be pressing reasons for taking this course of action. Generally, the practice should be discouraged because:

- The clerk of works is inspecting for the benefit of the employer, and owes no duty to the contractor to find defects

Figure 7.8
Letter from contractor to architect if the clerk of works defaces work or materials

Dear Sir

PROJECT TITLE

It is common practice for the clerk of works to deface work or materials considered to be defective. The basis for such action is to bring the defect to the notice of the contractor and ensure that it cannot remain without attention.

We object to the practice because:

(1) The work or materials so marked may not be defective and we will be involved in extra work and the employer in extra cost in such circumstances.
(2) The work or materials so marked, if indeed defective, will not be paid for and will be our property when removed. We may be able to incorporate it in other projects where a different standard is required. Defacement by the clerk of works would prevent such reuse.

We will take no action about the defacing marks we noted on site today, but if the practice continues, we will seek financial reimbursement on every occasion.

Yours faithfully

- It is the job of the person in charge to produce lists of defective work
- The contractor may be under the impression, however misguided, that if it simply attends to defects on the 'snagging lists', its obligations are at an end, and disputes may follow.

Obviously, the clerk of works must draw the contractor's attention to work not in accordance with the contract documents, but should not be more specific. In particular, neither the clerk of works nor the architect should instruct how defective work is to be corrected if the only problem is that it is not in accordance with the contract. To do otherwise may result in the employer having to pay for the work: *Simplex Concrete Piles Ltd* v. *Borough of St Pancras* (1958).

7.2.3 Responsibilities

Like everyone else connected with the contract, the clerk of works has a responsibility to carry out relevant duties in a competent manner. The clerk of works must demonstrate the same degree of skill that would be demonstrated by the average clerk of works. If a clerk of works purports to be especially skilled in some branch of work, a greater standard of skill than the average will be expected in that particular branch.

Case law has made clear that, provided the ordinary relationship of master and servant exists between employer and clerk of works, the employer will be vicariously liable for the actions of the clerk of works in the normal way: *Kensington and Chelsea and Westminster Area Health Authority* v. *Wettern Composites and Ors* (1984); *Gray* v. *T. P. Bennett & Son* (1987). The fact that the clerk of works is under the architect's direction makes no difference. The relationship between the clerk of works and architect has been compared to that between the chief petty officer and the captain of a ship. This does not mean that, if the clerk of works is negligent, it will relieve the architect of all responsibility, but it may substantially reduce the architect's liability for damages depending on circumstances. It is, therefore, very important to ensure that the employer employs a clerk of works. The negligence of a clerk of works employed by the architect will not reduce the liability of the architect.

7.3 *Summary*

The quantity surveyor

- A suitable clause can be included in the contract to provide for the appointment of a quantity surveyor and for appropriate duties.

- It is for the architect to advise the employer on the appointment.
- There is a danger in the architect purporting to carry out quantity surveying functions.
- An architect acting as quantity surveyor must check with the professional indemnity insurers.
- A clause should be inserted in the contract or a letter written to the contractor to cover the quantity surveyor's duties.
- The quantity surveyor's duties should include deciding quantum, but not liability to pay.
- The quantity surveyor should be appointed by, and be liable to, the employer.
- It is the architect's responsibility to check that all certificates are correct.
- Despite legal liabilities, the employer will probably look to the architect first if something goes wrong.

The clerk of works

- No provision in the contract.
- The architect should advise the employer whether a clerk of works is necessary.
- A suitable clause should be inserted in the contract.
- The clerk of works should be appointed by the employer.
- The clerk of works should be appointed immediately the successful tender has been accepted.
- The clerk of works should be briefed thoroughly.
- Sole duty should be to inspect the Works.
- May carry out other duties for the architect.
- Must not put marks on the work.
- Should not issue 'snagging lists'.
- Should not instruct how defective work is to be made good in accordance with the contract.
- Must be competent.
- The employment of a clerk of works may reduce the architect's liability for damages if the employer is found to be vicariously liable.

CHAPTER EIGHT
SUBCONTRACTORS AND SUPPLIERS

8.1 *General*

This chapter deals with third parties insofar as they are provided for or might affect the work carried out under MW and MWD. Subcontractors, suppliers, statutory authorities, persons engaged directly by the employer and the possibility of nominating or naming subcontractors are considered.

The contract provisions are extremely brief. They are contained in clauses 2.1, 2.5 (2.6 in MWD), 3.1 and 3.3. There are no provisions for suppliers, employer's licensees or nominated or named subcontractors.

8.2 *Subcontractors*

8.2.1 Assignment

Assignment is usually coupled with subcontracting in contract provisions and MW and MWD are no exception. It is a mistake, however, to consider that they are linked. They are totally different concepts. Assignment is the legal transfer of a right or duty from one party to another whereby the original party retains no interest in the right or duty thereafter. For this to be fully effective *novation* must take place, that is, the formation of a new contract.

Subcontracting is, in essence, the delegation of a duty from one party to another, but the original party still retains primary responsibility for the discharge of that duty. It is vicarious performance of a duty by someone else.

Clause 3.1 deals with assignment. Both employer and contractor are prohibited from assigning the contract unless one has the written consent of the other. This is a much stricter provision than is to be found under the general law where it is possible for either party to assign the benefits or rights of a contract to a third party. For example, it is quite common for the contractor to wish to assign to a third party the benefit of receiving progress payments in return for substantial financial help at the beginning of the contract. The employer, too, may wish to sell an interest in the completed building to another before the issue of the final certificate, thus assigning the benefit of the contract. Although permitted under the general law, clause 3.1 operates to stop such deals. Duties or burdens of contracts can never be assigned without express agreement between the parties. The effectiveness of this kind of clause was considered by the House of Lords in *Linden Gardens* v. *Lenesta Sludge Disposals* and *St Martin's Property Corporation* v. *Sir Robert McAlpine & Sons* (1993). The clause was held to be effective to prevent assignment of the benefits of a contract or the right to damages for breach of the contract.

It is considered that the architect has a duty to explain the possible difficulties to the employer, particularly if there is reason to believe that the employer might wish to assign a benefit before the final certificate is issued. Assignment can be made with the written consent of the other party, but such consent may be refused and the grounds for refusal need not be reasonable. It would appear to do no harm, and make much sense, for the architect to advise the employer to amend the contract so as to prohibit only the assignment of duties under the contract. The amendment should be carried out at tender stage and should result in lower rather than higher tenders.

8.2.2 Subcontracting

Clause 3.3 deals with subcontracting. Subcontracting is traditional in the building industry, but the practice can be abused to the extent that the contractor's sole employee on the site might be the person-in-charge, the remainder of the workforce being subcontractors. Needless to say, such an arrangement does not make for an efficient contract. Clause 3.3.1 is designed to prevent this and other problem situations by allowing subcontracting only if the architect gives written consent (see Figure 8.1). Although the architect may withhold consent, he or she must act reasonably. It is probably reasonable for the architect to require the contractor to supply the name of the proposed subcontractor before consent is given.

It is important to remember that there is no contractual relationship between the employer and the subcontractor. The subcontractor's contract is with the contractor. It follows that nothing contained in MW or MWD is binding in any way on the subcontractor. It is vital, therefore, that the subcontract gives the contractor sufficient controls over the subcontractor because the employer must look to the contractor for redress if the subcontractor defaults.

Figure 8.1
Letter from contractor to architect requesting consent to subletting

Dear Sir

PROJECT TITLE

We propose to sublet portions of the works as indicated below because [*state reasons*]. We should be pleased to receive your consent in accordance with clause 3.3.1.

[*list the portions of the works and the name of the subcontractor*]

Yours faithfully

There is no standard form of subcontract for use with MW or MWD. It is thought that the architect would not be unreasonable to withhold consent to subletting until satisfied that the contractor's subcontract provisions were adequate. The architect should beware, however, of being drawn into disputes between contractor and subcontractor or of being seen to approve the form of subcontract. It should be remembered that ultimately the contractor is responsible for its own subcontractors and it has no right to look to the architect or the employer to assist if things go wrong.

Clause 3.3.1 introduces a term which confirms that, despite any subcontracting, the contractor is wholly responsible for carrying out and completing the Works. In light of the meaning of subcontracting, it is questionable whether this term is strictly necessary.

The lack of detailed provisions is sometimes to be regretted, but is unavoidable taking the contract as a whole. Unfortunately, although the number of subcontractors will tend to increase with the size of the work, there is no lower point at which the contractor will not use any subcontractors at all. Clause 3.3.2 stipulates that a subcontract must include a provision which makes the contractor liable to pay simple interest to a subcontractor if the contractor fails to pay the amount due to the subcontractor by the final date for payment. The rate of interest is to be 5% over the Bank of England Base Rate as in clause 4.4 where late payment to the contractor is concerned. The interest is to be treated as a debt owing from the contractor to the subcontractor.

Clause 3.3.3 provides that the subcontractor's employment under the subcontract is to terminate immediately on the termination of the contractor's employment under the main contract. This provision will not affect the subcontractor, of course, unless it is incorporated in the subcontract. The positioning of this clause may be a drafting error, because the clause immediately follows clause 3.3.2 which does require the contractor to incorporate a term regarding interest into the subcontract. It seems possible that the opening of clause 3.3.2 was intended to embrace clause 3.3.3 also. Clause 3.3.2 does not impose any further subcontract terms on the contractor.

If the architect does decide to check the contractor's form of subcontract and it is not a recognised form, Figure 8.2 shows some of the matters for which it should make provision. Some of them are now imposed by the Act in any event. It is essential that the subcontract steps-down the relevant main contract provisions. The architect has no responsibility for the subcontract and any checking will be done merely to ensure that the contractor has avoided situations which might have repercussions on

Figure 8.2
Subcontract checklist

- Subcontractor's obligations
- Information: supply and timing of documents, drawings and details; custody and confidentiality; design liability
- Instructions, variations and valuations
- Access to site for subcontractor; access to subcontract works for architect; regulations
- Non-exclusive possession of the site
- Assignment and subcontracting
- Vesting of property and insurance
- Commencement, completion and extension of the subcontract period
- Completion date and making good of defects
- Payment in instalments*
- Withholding of money*
- Suspension of obligations*
- Financial claims
- Damages for delay
- Determination
- Construction Industry Scheme
- Disputes procedure – adjudication,* arbitration

*Clauses imposed by the Housing Grants, Construction and Regeneration Act 1996.

the main contract. Any dispute between contractor and subcontractor is potentially disruptive.

8.2.3 Nominated subcontractors

MW and MWD make no provision for nominated subcontractors. Their use for works for which this form would be suitable is not envisaged. Having said that, there are various devices which can be used to provide for 'nominated' subcontractors if the architect or the employer is absolutely sure that they must 'nominate' a particular firm.

The widespread use of nominations in building contracts has come to be associated with certain laziness or shortage of time on the part of the architect. It often seems quicker and less trouble in the short term to put a sum in the bills of quantities or specification than to properly specify requirements at the outset. It is obviously a temporary expedient only and at some future date, usually sooner rather than later, the architect is faced with specifying the work in question and engaging in the complicated process of nomination. Difficult legal and administrative problems can result during the life of the contract – as numerous legal cases testify. The message is clear. Wherever possible, nomination should be avoided. If nomination is required consideration should be given to using IC or ICD, which provide for 'naming' subcontractors in a clause of complexity and some obscurity, or ACA 3, which has relatively simple 'naming' procedures. If 'nomination' or something similar is desired using MW or MWD, it is possible to use one of the following methods:

- Name one or a choice from several named firms in the specification
- Name a firm in an instruction directing the expenditure of a provisional sum in accordance with clause 3.7
- Include an appropriately worded clause in the contract
- Provide for the specialist firm to be directly employed by the employer.

There are severe pitfalls inherent in the use of any of these methods and the architect should seek the advice of an experienced construction contract consultant before proceeding. Among the points to be considered are the following.

Naming in the specification

It is perfectly possible for the architect to name one firm in the specification which the contractor must use for carrying out a specific part of the work. If this is done, a considerable amount of power is placed in the hands of such a firm which can, in effect, quote to the contractor whatever price it likes, secure in the knowledge what whichever contractor is awarded the contract, the specialist firm will be incorporated. If, as is not unknown, the specialist firm goes into liquidation before work begins on site, the architect has an instant problem with which to start contract administration duties: whether that is sufficient to amount to frustration of the contract is possibly debatable. In such circumstances, the architect will have to move quickly to issue an instruction varying the name of the subcontractor, perhaps doing so more than once to deal with possible objections from the contractor.

The biggest difficulty with this type of nomination is that there is no provision in the contract to deal with the consequences. The precise extent of those consequences can be envisaged by glancing through clause 35 of JCT 98. The answer, of course, is to incorporate a fairly substantial clause in the contract to deal with the situation. Such a clause would require very careful drafting.

The alternative, whereby the contractor is given a choice of three or four names in the specification, is not strictly nomination at all. However, it does give a degree of control over the firms to be used, while allowing the contractor to seek competitive quotations. A subcontractor appointed by the contractor under this method would be the contractor's entire responsibility to the extent that, if it failed, it would be the contractor's job to find an alternative. Although the worst consequences of nomination can be avoided by this system, it is advisable to include a suitably worded clause in the contract.

Naming by an instruction in respect of a provisional sum

If it is intended to operate this system, it is essential that the contractor knows what is intended at the time of tender so that it can make suitable provision in its price. A big danger is that the contractor might strongly object to the firm in question when it is named for the first time in an instruction. There is no machinery to deal with its objection and the efficient progress of the work can be seriously affected. All the comments regarding the naming of a single firm in the specification are equally applicable to this case and an amendment to clauses 3.3 and 3.7 is indicated.

Including a clause in the contract

The trouble with this is that once substantial clauses are added dealing with particular circumstances, in this case nomination, there is the risk of the contract being considered the employer's 'standard written terms of business', which brings it within the scope of the Unfair Contract Terms Act 1977. The clauses are also liable to be interpreted against the employer if there is any ambiguity. There is no doubt that a suitable clause can be formulated and, if nomination is what is wanted, it is probably the safest way to achieve it. Nomination can proceed along time-honoured procedures and potential difficulties can be provided for. On the other hand, to add a nomination clause like clause 35 in JCT 98 would effectively double the size of the contract – to say nothing of the ancillary documentation which would be necessary.

The employer directly employing the specialist firm

This may seem an attractive solution, but there are dangers, not least the problem of integration with the Works (see section 8.4).

8.3 Statutory authorities

MW and MWD do not expressly mention statutory authorities. However, clause 2.1 states that the contractor must comply with statutory requirements, which are defined in clause 1.1 to include any statute, any statutory instrument, rule or order or any regulation or bye-law applicable to the Works. This means, for example, that the contractor must not contravene the Planning Acts or regulations made in pursuance thereof, and must comply with the Building Regulations and, of course, the Construction (Design and Management) Regulations 1994. By necessary implication, therefore, the contractor must allow statutory undertakers to enter the site and carry out work which they alone are empowered to do.

Certain crucial parts of virtually all contracts are carried out by statutory undertakers such as local authorities, gas, water and electricity suppliers. When they carry out work solely as a result of their statutory rights or obligations, they are in a special category quite separate from subcontractors or employers' directly employed firms. If completion of the Works is delayed as a result of a supplier carrying out or failing to carry out work in pursuance of its statutory duty, the contractor will be entitled to an extension of time in accordance with clause 2.7 (clause 2.8 in MWD).

If the contractor employs a statutory undertaking to carry out work which is not part of its statutory rights or duties, the statutory undertaking ranks as an ordinary subcontractor. Delay caused by the undertaking in such a case would not entitle the contractor to an extension of time under MW or MWD. An example should make the principle clear.

It is usual to include a provisional sum in the specification to cover the cost of connecting the electrical system of a dwelling to the mains. The contractor may also, with the architect's written permission, sublet the electrical wiring and fittings in the dwelling to the electricity supplier. The mains connection is part of the supplier's statutory obligations; the internal wiring and fittings are not part of the supplier's statutory obligations, but a matter of contract between the contractor and the supplier. If the mains connection delays the completion date, the contractor is entitled to an extension of time. If the internal wiring and fittings delay the completion date, the contractor is not entitled to any extension of time.

Statutory undertakers have no contractual liability when they are carrying out their statutory duties – a sore point with many architects – but in certain cases they have a liability in tort. Outside their statutory duties, they are in the same position as anyone else if they enter into a contract to carry out work.

The contractor must also give all notices required by statute etc. and must pay all fees and charges in respect of the works, provided that they are legally recoverable from the contractor, but not otherwise. The contractor is not entitled to be reimbursed and is deemed to have included the necessary amounts in the price. The exception to this, of course, is if the charge is a necessary result of an architect's instruction. In such a case, the amount of the charge will form part of the valuation of the instruction carried out under the provisions of clause 3.6.

Clause 2.5.2 (clause 2.6.2 in MWD) states that the contractor is not liable to the employer under the contract if the works do not comply with statutory requirements provided that:

- The contractor has carried out the works in accordance with the contract documents or any of the architect's instructions *and*
- If the contractor has found a divergence between the contract documents or the architect's instructions and statutory requirements, the contractor has immediately given him a written notice specifying the divergence.

The contractor is not liable if it fails to find a divergence which actually exists: *London Borough of Merton* v. *Stanley Hugh Leach Ltd* (1985). Although the contractor may be freed from liability to the employer, its duty to comply with statutory requirements remains. Thus, the local authority may serve notice on the contractor if work, built correctly in accordance with the contract documents, does not comply with the

Building Regulations. In such a case, the architect would have to act speedily to issue appropriate instructions if the employer was to avoid a substantial claim at common law for damages.

The contractor is not entitled to take any emergency measures to comply with statutory requirements even though delay might cost the employer money. The contractor's obligation to give the architect immediate written notification remains. If the emergency concerns part of the structure which is actually dangerous, the contractor has a responsibility under the general law to take whatever measures are necessary to make the structure safe. It is difficult to see how the contractor he could make any valid claim in such circumstances.

Clause 3.9 makes provision for compliance of the parties with the CDM Regulations. The idea is to make compliance with the regulations a contractual duty so that breach of the regulations is also a breach of contract. Clause 3.9.1 has been inserted to provide that the employer 'shall ensure' that the planning supervisor carries out all the relevant duties under the regulations and that, where the principal contractor is not the contractor, it will also carry out its duties in accordance with the regulations (clause 3.9.2). There are also provisions that the contractor, if it is the principal contractor, will comply with the regulations.

Every architect's instruction potentially carries a health and safety implication which should be addressed. The CDM Regulations pose a formidable list of duties on the planning supervisor. Some of these duties must be carried out before work is started on site. If necessary actions delay the issue of an architect's instruction or once issued delay its execution, the contractor will be entitled to extension of time and, depending on circumstances, it may have a common law claim for damages for breach of an express term of the contract. These are matters about which all parties should take great care.

8.4 *Works not forming part of the contract*

The contract makes no provision for the employer to enter into a contract with anyone other than the contractor to carry out any part of the work on site while the contract work is being carried out. It is quite common for the employer to wish to engage others to do certain work or, indeed, to use some of its own employees. The reason may be because the employer has a special relationship with the firm or individual, for example in the case of a sculptor, artist or landscaper, or because the employer wants complete control over a particular operation. When using MW or MWD, direct engagement by the employer can be achieved in one of two ways:

(1) With the consent of the contractor
(2) By including a special clause to that effect.

It is never a good idea to bring third parties onto a site during the contract period. Three dangers which merit special consideration are:

(1) It is easy for the contractor to claim that such persons have disrupted its work and/or delayed the completion date. They are very difficult claims to refute because introducing third parties clearly does not help the contractor and can usually be seen to cause at least some degree of hindrance. The contractor may be able to claim damages at common law and an extension of time under clause 2.7 (clause 2.8 in MWD).
(2) The contractor may acquire grounds to terminate its employment under the contract. Clause 6.8.2.2 provides for the contractor to terminate if the Works are suspended for a continuous period of one month because of any impediment, prevention or default of the employer, the architect or of any person for whom the employer is responsible. The employment of other contractors would certainly fall into this category. Whether the Works would be delayed by as much as a month would depend on the circumstances. This matter is dealt with in more detail in section 12.3.2.
(3) For insurance purposes, persons directly employed by the employer are deemed to be persons 'for whom the Employer is responsible' (clause 5.1). They are not deemed to be subcontractors. Therefore the employer may have uninsured liabilities. The directly employed persons may well have their own insurance cover, but it is the architect's duty to advise the employer to obtain the necessary cover through the employer's own broker. The cover should be for the employer and those persons for whom the employer is responsible in respect of acts or defaults occuring during the course of the work. It is a complex business and best left in the broker's hands. It is not the architect's responsibility to advise on insurance matters.

Statutory undertakers acting outside the confines of their statutory duties may be considered to be directly employed by the employer if they are not paid by the contractor and under its control. It is always prudent to ensure that the contractor is responsible for all the work to be carried out during the currency of a contract. It removes some possible areas of dispute and promotes efficiency. If the employer insists on having directly employed persons, remember that, for work to be considered as not forming part of the contract, it must:

- Be the subject of a separate contract between the employer and the person who is to provide the work *and*
- Be paid for by the employer direct to the person employed, not through the contractor.

8.5 Summary

Assignment

- A different concept from subcontracting
- Neither party may assign without the other's consent
- Under the general law, either party may assign benefits, but not duties
- It may be beneficial to amend clause 3.1.

Subcontracting

- The contractor may sublet with the architect's consent
- There is no contractual relationship between the employer and the subcontractor
- The subcontract terms may have important repercussions for the employer.

Nominated subcontractors

- No provision in MW or MWD
- If nominated subcontractors are required, a different contract form should be considered
- Devices can be used to enable nominations to be made when using MW or MWD, but there are pitfalls.

Statutory authorities

- The contractor must allow them to enter the site
- The contractor must comply with statutory requirements
- In pursuance of their statutory duties, they are not liable in contract, but may provide grounds for an extension of time
- Not in pursuance of their statutory duties, they are liable in contract and may be subcontractors or persons for whom the employer is responsible

- The contractor is not liable to the employer if it works to the contract documents, provided that it notifies the architect of any divergence it finds
- There is no provision for emergency work
- Compliance with the CDM Regulations is a contractual requirement
- Extension of time and other claims may arise from compliance with the CDM Regulations.

Work not forming part of the contract

- MW and MWD make no provision for directly employed persons
- The contractor must consent or a special clause must be included
- Fertile ground for claims
- Possible ground for determination
- Insurance implications
- Should be avoided.

CHAPTER NINE
POSSESSION, COMPLETION AND DEFECTS

9.1 *Possession*

9.1.1 Introduction

If there is no express term in the building contract, a term will always be implied that the contractor must have possession in sufficient time to allow it to finish the Works by the contract completion date: *Freeman* v. *Hensler* (1900). To have possession of something is the next best thing to ownership. If the owner of this book lends it to a friend, the friend can defend his or her claim to it against anyone except the original owner. The same principle applies to a contractor in exclusive possession of a site. It is in control of the site and has the power to refuse access to anyone else, including the employer. In practice, this stern rule is modified by the operation of numerous statutory regulations, allowing entry by the representatives of various statutory bodies, and by the express and implied terms of the contract.

In legal terms, the contractor is said to have a licence from the owner of the site to occupy it for the period of time necessary to carry out and complete the Works. The period of time is the period stated in the contract or any extended period. During the contractor's lawful occupation the employer has no power under the general law to revoke the licence, but the contract may contain express terms giving such power, for example, in the case of lawful termination of the contractor's employment. Case law suggests that the absence of such an express term may pose awkward problems if the contractor refuses to give up possession: *Hounslow Borough Council* v. *Twickenham Garden Developments* (1971). In general, however, if the contractor retains possession of the site after the contract period or any extended period has expired, it is in the position of a trespasser and can be removed under the general law.

This contract does not contain an express term requiring the contractor to give up possession immediately if the employer terminates the contractor's employment under the provisions of clauses 6.4, 6.5 or 6.6. However clause 6.7.1 allows the employer and any replacement contractor to take possession – which amounts to much the same thing. Although there is no similar specific requirement when the contractor terminates its own employment under clauses 6.8 or 6.9, a term to that effect will be implied.

There is no express provision for access to the Works for the architect equivalent to clause 3.1 of SBC, but such a term must be implied to allow the architect and authorised representatives to carry out their duties under the contract.

9.1.2 Date for possession

Clause 2.2 (clause 2.3 in MWD) relates to 'Commencement and completion', but it does not make reference to a date for possession. The contract particulars do, however, leave a space for a date to be inserted for 'commencement of the Works'. It will be implied that this date is the latest date on which the employer must give possession of the site to the contractor. The inclusion of the word 'may' in clauses 2.2 (clause 2.3 in MWD) could be significant. The straightforward interpretation of the clause appears to be that the date to be inserted in the contract particulars is the earliest on which the contractor will be allowed to commence carrying out the Works, but not necessarily the latest. In effect, it appears that the contractor would be within this clause if it started work in the fourth month of a six-month contract, provided that it finished by the contract completion date. Contrast with SBC, clause 2.4, where the contractor 'shall thereupon begin the construction of the Works . . . '. Of course, in practice, the contractor could well be in danger of having its employment terminated in accordance with clause 6.4.1.2, because that clause gives, as a ground for termination of the contractor's employment, failure to proceed regularly and diligently with the Works.

In *Greater London Council* v. *Cleveland Bridge & Engineering Co Ltd* (1986), the court was of the opinion that a requirement for the contractor to proceed with due diligence and expedition must be interpreted in the light of other requirements as to time. The contractor may, therefore, be able to argue that it is proceeding regularly and diligently provided it can meet the completion date and it is a moot point whether it could be said to be suspending something it had not yet begun. In any case, it might well argue that it had reasonable cause if, in fact, the contract period was very generous. 'Regularly and diligently' has now been comprehensively defined by the Court of Appeal: *West Faulkner* v. *London Borough of Newham* (1993). (See also section 5.1.2.)

If the employer fails to give the contractor possession so that it can begin the Works on the stated date, it will be a serious breach of contract. The contractor will have a claim for damages at common law and the completion date will cease to be operative: *Rapid Building Co Ltd* v. *Ealing Family Housing Association Ltd* (1985). In such circumstances the contractor's obligation would be to complete within a reasonable time. Although that does not mean that the contractor has unlimited time in which to carry out the Works, it does mean that there is no date from which liquidated damages can run and, therefore, they are not deductible. If the employer's failure to give possession lasts more than a few days, it seems that the contractor may well have grounds to consider the employer's breach as an intention to repudiate the contract. The contractor should serve notice on the employer (Figure 9.1). In any case, failure to give possession clearly requires negotiation between the employer and the contractor to achieve an amicable settlement and the architect should remind the employer of the duty to give possession before the due date (Figure 9.2) and give advice if the employer is in breach (Figure 9.3).

Although the architect's power to instruct the contractor probably extends to postponing the work, postponement is not the same as failure to give possession. If the architect postpones the work, the contractor will have possession of the site, but the carrying out of the Works will be suspended. The contractor may well wish to use the time to reorganise its site arrangements, repair site offices, improve its security, etc.

When the contractor has completed the Works, it normally gives up possession, but it has a restricted licence to continue to enter the site to deal with such defects as are notified to it under clause 2.10 (clause 2.11 in MWD), defects in the rectification period (see section 9.3).

9.2 *Practical completion*

9.2.1 Definition

Clause 2.2 (clause 2.3 in MWD) states that the Works must 'be completed by' a date to be inserted. The words have their ordinary meaning, that is to say the completion of the Works must not take place after the stated date, but it may take place before the stated date. Note,

Figure 9.1
Letter from contractor to employer if possession not given on the due date

SPECIAL OR RECORDED DELIVERY

Dear Sir

PROJECT TITLE

Possession of the site should have been given to us on [*insert date*] to enable us to commence the Works on [*repeat date*] in accordance with clause 2.2 [*substitute '2.3' when using MWD*] of the conditions of contract. Possession was not given to us on the due date.

This is a serious breach of contract for which we will require appropriate compensation and we reserve all our rights and remedies in this matter. Without prejudice to the foregoing, we suggest that a meeting would be useful and look forward to hearing from you.

Yours faithfully

Figure 9.2
Letter from architect to employer before date for possession

Dear Sir

PROJECT TITLE

The contractor is entitled, by the terms of the contract, to take possession of the site on the [*insert date*].

Will you be certain that everything is ready so that the contractor can take possession? Failure to give possession on the due date is a serious breach of contract which cannot be remedied by a simple extension of the contract period; the contractor may be able to claim substantial damages or even treat the contract as repudiated.

Please let me know immediately if you anticipate any difficulties.

Yours faithfully

Figure 9.3
Letter from architect to employer if there is a failure to give possession on the due date

Dear Sir

PROJECT TITLE

I understand/have been notified by the contractor [*delete as appropriate*] that you were unable to give possession of the site on the date stated in the contract as the date on which the contractor may commence the Works.

You will recall that in my letter of the [*insert date*] I pointed out that failure to give possession is a serious matter.

It is something which you must negotiate with the contractor if you wish to avoid the charge of repudiation and heavy damages. With your agreement, I will try to negotiate on your behalf but, since this is outside the terms of my appointment, I should be pleased to have your written authorisation to act for you in this way and your agreement to pay my additional fees and costs on a time basis as laid down in my original conditions of appointment.

Yours faithfully

however, that there is no provision for partial possession by the employer. This omission is entirely reasonable in view of the small-scale nature of the Works for which this contract is intended to be used and the correspondingly short time-scale. It is extremely unlikely that the employer will need to take possession of part of the Works before completion. If it is decided that provision must be made for partial possession it is worth considering the use of IC or ICD instead. If phased completion is to be provided, MW and MWD are not suitable and either SBC, IC, ICD or ACA 3 would be appropriate. It is not envisaged that partial possession, much less sectional completion, would be a feature of work for which MW or MWD were being considered.

Clause 2.9 (clause 2.10 in MWD) states that the architect must certify the date when in his or her opinion the Works have reached practical completion and the contractor has complied sufficiently with clause 3.9.3. Clause 3.9.3, of course, requires the contractor to provide and to ensure that any subcontractor provides information for the preparation of the health and safety file required by the CDM Regulations. Therefore, the architect is to certify the date by which both criteria have been satisfied. Only one date may be certified and not, as some architects believe, one date for each criterion. In accordance with clause 2.3 (clause 2.4 in MWD), the architect must issue the certificate in writing. Although no time scale is indicated, the architect must issue the certificate within a reasonable time of the certified date because it is a particularly important stage in the contract (see section 9.2.2). In practice, the certificate should be issued immediately.

Note that it is the architect's opinion which is required by the contract, not that of the employer or the contractor. Despite the significant consequences of the architect's certificate, 'practical completion' is nowhere defined in the contract. It does not mean substantially or almost complete and the precise meaning has exercised several judicial minds. On balance, it seems to mean the stage at which there are no defects apparent and only very trifling items remain outstanding: *H. W. Nevill (Sunblest) Ltd* v. *Wm Press & Son Ltd* (1981). What qualifies as 'trifling' will depend on circumstances. It is not thought that the architect is justified in withholding the certificate until every last screw and spot of paint is in place. That would indeed be completion, but the contract clearly intends something rather short of that by the use of the word 'practical'. Within these guidelines the architect is free to exercise some discretion. The architect would not be justified in issuing a certificate, in any event, if items remained to be finished which would seriously inconvenience the employer. Note that the architect cannot issue the certificate even when in the architect's opinion practical completion has been achieved.

The CDM provisions must also be satisfied. (See above and the effect on the deduction of liquidated damages in section 10.2.2.)

The architect is under no obligation to issue lists of outstanding items if the certificate is withheld. Clerks of works often consider it part of their duty to supply the contractor with so-called 'snagging lists'. It is a bad practice because the contractor tends to assume that when it has completed the lists, its obligations are at an end; therefore, disputes sometimes occur. The contractor's obligations should be clear from the contract documents and the duty to make sure that the work is complete in accordance with the contract lies with the contractor, not with the architect or the clerk of works (if employed). More particularly, any 'snagging lists' should be prepared by the contractor's person in charge as part of the normal supervision of the Works. Whether that is done is not the architect's direct concern.

The architect's duty to issue a practical completion certificate does not depend on any request by the contractor. The duty must be carried out as soon as the architect is satisfied that the criteria have been satisfied. Many architects arrange a hand-over meeting to which the employer and any consultants are invited. It should be remembered, however, that the architect cannot transfer responsibility for certifying practical completion to the employer, although it may be a prudent move to see that the employer is happy with the building before taking possession. In practice, it is much more likely that the employer will wish to take possession before the architect is thoroughly satisfied. In such circumstances the architect must strictly observe his or her duty and refuse to issue the certificate until so satisfied. The contractor will then gain no benefit and may be at some disadvantage in completing the work. The contractor may complain to the employer who may, in turn, complain to the architect who should put the position to the employer in writing for the record (Figure 9.4). If the architect submits to pressure, it may well leave open a future claim for negligence, not only from the employer but also from third parties to whom the architect may have given a warranty that duties will be carried out with reasonable skill and care.

A decision in the Technology and Construction Court (*Skanska Construction (Regions)* v. *Anglo-Amsterdam Corp* (2002)) at first sight appears to hold that if the employer takes possession of the building, practical completion is deemed to have occurred. The decision should be read with care. The JCT form in question was amended to provide a stringent test for practical completion and, after possession by the employer, the contractor was allowed back only with permission. Nevertheless, to deem practical completion when it had not in fact occurred appears perverse and it is difficult to understand the grounds for the decision.

Figure 9.4
Letter from architect to employer if possession of the Works has been taken before practical completion

Dear Sir

PROJECT TITLE

I refer to your letter of the [*insert date*].

I confirm that, in my opinion, practical completion has not been achieved. Therefore, it is my duty, about which I have no discretion, to withhold my certificate. I note, however, that you have agreed with the contractor to take possession of the building. I think this is unwise, but I will continue to inspect until I feel able to issue my certificate. At that date, the rectification period will commence. Naturally, the contractor has a very great interest in obtaining a certificate of practical completion and you must expect him to continue to complain until, in my opinion, practical completion is achieved. You should be aware that taking possession before practical completion may give rise to problems. For example, the insurance position will be unclear and you should seek advice from your broker. In addition, the contractor may seek additional payment, because it is now having to work in an occupied building.

Yours faithfully

The better view is that occupation by the employer is not equivalent to practical completion: *BFI Group of Companies Ltd* v. *DCB Integrated Systems Ltd* (1987); *Impresa Castelli SpA* v. *Cola Holdings Ltd* (2002).

9.2.2 Consequences of practical completion

The issue of the certificate of practical completion is of particular importance to the contractor because it marks the date when:

- The rectification period commences (clauses 2.10 and 2.11 in MW and MWD respectively)
- The contractor's liability for liquidated damages ends (clauses 2.8.1 and 2.9.1 in MW and MWD respectively)
- The employer's right to deduct full retention ends and half the retention held becomes due for release within 14 days (clause 4.5)
- The machinery culminating in the issue of the final certificate is set in motion (clause 4.8)
- The contractor's liability to insure under clause 5.4A ends.

9.3 *Rectification period*

9.3.1 Definition

The rectification period is inserted for the benefit of both parties. It allows a period of time for defects to appear and to be corrected with the minimum of fuss. Any defect which is the fault of the contractor is a breach of contract on the part of the contractor and, without such a period, the employer would have no contractual remedy. The employer would be left to common law rights. Moreover and more importantly, if there were no rectification period, the contractor would have no right to re-enter the site to remedy the defects. If the employer suffers some loss as a direct result of the defects and the mere remedying of those defects is not adequate restitution, action at common law is always available to obtain damages from the contractor for the breach.

Contractors commonly hold two mistaken views about the rectification period:

(1) That the contractor's liability for remedying defects ends at the end of the rectification period
(2) That the contractor is liable to correct anything which is showing signs of distress or with which the employer is not satisfied.

With regard to the first, the contractor's liability does not end at the end of the rectification period. What does end is its privilege to return and remedy the breach. After that time, the contractor remains liable, but after proper notice the employer can simply pursue an action at common law for damages, if so desired, without giving the contractor the opportunity to return.

The second mistaken view probably owes its origin to the practice of referring to the rectification period as the 'maintenance period'. Architects and contractors alike are guilty in this respect (even GC/Works/1 (1998) and ACA 3 use the term) although maintenance implies a heavier responsibility than simply making good defects. Repolishing, cleaning and generally keeping a building in pristine condition could be said to be maintenance for which the contractor has no responsibility. The employer's dissatisfaction with the building is also of little consequence in itself (see section 9.3.2) if the contractor has carried out its obligations. If instructed to carry out what amounts to matters of routine maintenance, the contractor should write to the architect making the position clear (see Figure 9.5).

9.3.2 Defects, shrinkages and other faults

Clause 2.10 (clause 2.11 in MWD) requires the contractor to make good defects, shrinkages and other faults. Case law has established that the phrase 'other faults' must be interpreted *ejusdem generis* with defects and shrinkages; that is to say, faults of the same kind. Read in this way 'other faults' appears to add little if anything to the contractor's liability. A defect can only mean something which is not in accordance with the contract. If the employer is unhappy about the paintwork, it could be that the contractor has not applied it correctly in accordance with the specifications. On the other hand, it could be that the specification is inadequate. Only the former situation would give rise to liability on the part of the contractor. An inadequate specification is usually the architect's responsibility.

Under MW 98 not all instances of shrinkage fell within the defects liability clause. Shrinkages had to be excessive. The intention appeared to be to exclude those shrinkages which could be said to be an unavoidable consequence of building operations. This was an eminently sensible approach in principle, but one which could result in a dispute because what was excessive to the architect may have been trifling to the contractor. Indeed, the whole question of shrinkages is fraught with difficulty. They

Figure 9.5
Letter from contractor to architect regarding routine maintenance

Dear Sir

PROJECT TITLE

We have received your instructions dated [*insert date*] which you purport to issue under the provisions of clause 2.10 [*substitute '2.11' when using MWD*] of the conditions of contract. Among the items listed as defects for us to make good are [*list defects complained of*].

Clause 2.10 [*substitute '2.11' when using MWD*] requires us to make good defects, shrinkages or other faults provided that, among other things, they are due to materials and workmanship not in accordance with the contract. In this instance, the materials and workmanship are demonstrably in accordance with the contract. What you are in effect instructing us to do is to carry out items of routine maintenance. Such is not our obligation under the contract and you have no power to issue such an instruction.

Yours faithfully

are the contractor's liability only if they result from workmanship or materials which are not in accordance with the contract. In practice, since the employer holds the purse-strings, it is the contractor who has to convince the architect that the shrinkage is not its responsibility. Shrinkage usually occurs due to loss of moisture or thermal movement. A common example is the shrinkage which occurs in timber after the building is heated. The architect will have specified a maximum moisture content for the timber, and subsequent shrinkage can only be because the timber was installed with, or allowed to develop, too high a moisture content, or the architect's specification was wrong. It is a question of fact rather than law, but note that it is no defence for the contractor to say that the architect's specified moisture content was impossible to achieve under normal site conditions. It is well known that it is difficult to maintain a low moisture content under the normally damp conditions which prevail on site, but that is not to say that such conditions cannot be improved by the use of suitable temporary heaters, ventilation, etc. The contractor's obligation is to provide workmanship and materials in accordance with the contract and it would have been well aware of the architect's requirements at tender stage. What it is really saying, therefore, is that it found it too expensive to comply with the architect's specification, and that is no defence at all.

The 'excessive' qualification to 'shrinkages' has been dropped for the MW and MWD contracts and, therefore, all shrinkages are now included provided only that they result from materials and workmanship not being in accordance with the contract.

9.3.3 Frost

The contractor's liability to make good frost damage is no longer expressly stated but in any event it would be limited to damage caused by frost which occurred before practical completion. This is perfectly reasonable since the contractor was in control of the Works up to, but not after, practical completion. Damage due to frost occurring after practical completion is the responsibility of the employer. In practice, there should be no great difficulty in detecting the difference. Frost damage after practical completion may be due, for example, to faulty detailing, unsuitable materials or lack of proper care by the employer. Note that the test is not when the damage occurred, but when the frost, which resulted in such damage, occurred.

9.3.4 Procedure

The rectification period starts on the date of practical completion as stated in the architect's certificate. If no period of time is inserted in the contract particulars, the period will be three months. There is really no good reason for limiting the period to three months because there is no connection between the length of the contract period and the rectification period. Whether the contract period is long or short, there will be some items of work completed just before practical completion and it is these items which the architect must consider when advising the employer of the length of rectification period required in a particular case. In general, there is probably much to be said for always inserting twelve months as the length of period on the basis that the building will be tested against all four seasons of the year. It is probably true that the contractor will include a slightly increased tender figure if twelve months is included as the rectification period, but there is really no reason why it should do so. It probably stems from a mistaken idea of the limits of its liability (see section 9.3.1). The final certificate will, of course, be delayed, but only $2\frac{1}{2}\%$ of the retention will be outstanding and payment will then probably closely follow the architect's certificate that all defects have been made good (see Chapter 11).

The defects etc. which the contractor is to make good are those which 'appear within' the rectification period. The wording suggests that any defects which have already appeared before practical completion could not be included as defects which the contractor must make good. In practice, the situation is not as bad as that. Defects which were apparent (sometimes referred to as 'patent defects') before practical completion would preclude the architect from issuing the certificate of practical completion (see section 9.2.1). Moreover, since no certificate is conclusive under this contract, the contractor's obligation to carry out the work in accordance with the contract is not reduced by the issue of any such certificate and if the contractor is so misguided as to refuse the opportunity of remedying the defects during the rectification period, the employer retains the normal common law rights intact. The danger is that if the architect certifies practical completion while there are some patent defects, half the retention will be released and the contractor may never return. The employer's common law rights will be useless if the contractor has gone into liquidation and the employer may, rightly, say that the architect was negligent in the issue of the certificate and look to the architect for recompense. If the architect does overlook some defects before issuing the certificate, the sensible thing to do is to include them with the defects which actually appear 'within' the period. This may not be precisely what the contract says, but it does no violence to the contractor's rights. This approach was noted with approval in *William Tomkinson* v. *Parochial Church Council of St Michael* (1990).

There is a requirement for the architect to notify the contractor about the defects. It is not expressly stated that the notice must be in writing. In practice, a written notice is required and clause 2.3 (clause 2.4 in MWD) probably covers the situation. There is no reference to the timing of such notice and it is thought that the normal practice of notifying the contractor at the end of the period would be acceptable. It is prudent to organise an inspection a few days before the period expires so that the architect can issue the final list of defects on the final day. The contractor should be prompt to respond if some items are not its responsibility (see Figure 9.6). If there is an urgent defect, such as a burst pipe or leaking roof, clause 2.10 (clause 2.11 in MWD) would entitle the architect to notify the contractor about such defect before the end of the period. In any event, it is thought that the architect can use the powers under clause 3.4 to instruct the contractor to carry out urgent remedial action during the rectification period.

There is no time limit set for the contractor to make good the defects, but the architect is not obliged to certify that making good has been achieved until satisfied (to do otherwise would be negligent) and the issue of the final certificate (clause 4.8.1) is dependent upon the architect's certificate under clause 2.11 (clause 2.12 in MWD) having been already issued. The contractor must carry out its obligations within a reasonable time although that is not expressly stipulated. What is reasonable will depend on:

- The number and type of defects
- Any special arrangements to be made with the employer with regard to access.

If the contractor fails to carry out its obligations, the architect may put the compliance procedure under clause 3.5 into operation (see section 4.3). If the contractor still fails to make good it should be noted that the employer may have the defects made good by others and all the costs involved may be deducted from the contract sum. All defects are to be made good by the contractor entirely at its own cost unless the architect instructs otherwise. Some commentators have been greatly concerned by this last phrase, even going so far as to suggest that it can only mean that the architect can instruct the contractor to remedy defects at the employer's expense. It is suggested that such a singular view is nonsense.

Figure 9.6
Letter from contractor to architect after receipt of schedule of defects

Dear Sir

PROJECT TITLE

Thank you for your instruction number [*insert number*] dated [*insert date*] scheduling the defects you require making good now that the rectification period has ended.

We have carried out a preliminary inspection and we are making arrangements to make good most of the items on your schedule. However, we do not consider that the following items are our responsibility for the following reasons:

[*list, giving reasons*]

We shall, of course, be happy to attend to such items if you will let us have your written agreement to pay us daywork rates for the work.

Yours faithfully

There is no doubt that the wording could be clearer, but it covers two distinct situations which might arise:

(1) If the defects are partly the fault of the contractor and partly contributed to by some default of the architect or the employer. In such circumstances it would be wrong to expect the contractor to remedy the defects entirely at its own expense and the architect might wish to direct the contractor regarding the extent to which it would be expected to bear the cost. The key words in the clause are 'entirely' and 'unless'.

(2) If the employer does not wish the contractor to remedy certain defects because, for example, to do so would seriously disrupt the employer's business. In such circumstances, the architect might wish to instruct the contractor not to make good such defects.

In either of the above situations, the architect must be sure to discuss the matter thoroughly with the employer before issuing any instruction. In the second case, the architect must obtain a letter from the employer authorising an instruction to the contractor that making good is not required (Figure 9.7). If the architect so instructs, note that there is no provision for any deduction from the contract sum. However, it appears that the architect in such circumstances would be entitled to reduce the contract sum by the amount it would have cost the contractor to make good: *William Tomkinson* v. *Parochial Church Council of St Michael* (1990).

Where defects appear after the end of the rectification period, the contractor is, of course, still liable, because each defect is a breach of contract. The contractor must be notified (*London & S.W. Railway* v. *Flower* (1875), but the employer is not obliged to request the contractor to make good and the contractor is not obliged to respond to such a request although it may be sensible to do so. Otherwise, the employer is entitled to engage others to make good and recover all the costs from the contractor: *Pearce & High* v. *John P. Baxter & Mrs A. Baxter* (1999).

Although clause 2.10 and 2.11 in MWD do also empower the architect (as other commentators suggest) to instruct the contractor to make good defects at the employer's expense, it is not something which the architect should ever consider doing. When making good has been completed, the architect must issue a certificate to that effect. The certificate has important implications with regard to the issue of the final certificate. Essentially, the final certificate cannot be issued unless the clause 2.11 (clause 2.12 in MWD) certificate has already been issued.

Figure 9.7
Letter from architect to employer if some defects are not to be made good

Dear Sir

PROJECT TITLE

I understand that you do not require the contractor to make good the following defects:

[*list*].

These defects are included in my schedule of defects issued at the end of the rectification period. In order that I may issue the appropriate instructions in accordance with clause 2.10 [*substitute '2.11' when using MWD*], I should be pleased if you would confirm the following:

(1) You do not require the contractor to carry out making good to the defects listed in this letter.
(2) You waive any rights you may have against any persons in regard to the items listed as defects in the above-mentioned schedule of defects and not made good.
(3) You agree to indemnify me against any claims made by third parties in respect of such defects.

Yours faithfully

9.4 *Summary*

Possession

- Must be given to the contractor in sufficient time for it to complete the Works
- Cannot be revoked by the employer under the general law, but the contract may give such power
- Is the same as the date on which Works may be commenced in MW and MWD
- Is the contractor's right, and failure on the part of the employer is a serious breach of contract
- Ends at practical completion.

Practical completion

- Is a matter for the architect's opinion alone subject to case law
- Requires the architect to issue a certificate; there is no provision for sectional completion or partial possession
- Bestows important benefits on the contractor.

Rectification period

- Benefits the contractor
- The end of the period does not mean the end of the contractor's liability for defects
- Defects are severely limited in scope
- May be of any length
- Is not a maintenance period
- Only those defects appearing within the period are covered under the contract
- Defects must be notified to the contractor during or at the end of the period
- Defects must be made good at the contractor's own cost
- The architect has power to instruct the contractor not to make good some defects or to instruct what proportion of cost must be borne by the contractor; check with the employer first
- When making good is complete, the architect must issue a certificate to that effect.
- Subject to the limitation period, the contractor is responsible for all the costs of making good defects which appear after the end of the rectification period even if the employer engages others to make good.

CHAPTER TEN
CLAIMS

10.1 General

It is traditional in the construction industry for claims by the contractor for both extra time and extra money to be linked together. However, it should be borne in mind that there is no necessary link between time and money. The grant of an extension of time is not a pre-condition to a monetary claim for loss and/or expense under any of the JCT Forms, neither does an extension of time automatically entitle the contractor to loss and/or expense. There can be money claims for both prolongation and disruption. MW and MWD are unique among the JCT contracts in that they contain no clause entitling the contractor to direct loss and/or expense except for the provision in clause 3.6 requiring the valuation of variations to include any direct loss and/or expense incurred by the contractor due to regular progress of the Works being affected by compliance with a variation instruction or due to compliance or non-compliance by the employer with clause 3.9 (obligations in regard to the planning supervisor and the principal contractor).

But the absence of such a clause should not mislead either employer or architect into thinking that the contractor cannot make financial claims. It can; but such claims must be pursued at common law by way of adjudication, arbitration or litigation and must be based on breach of some express or implied term of the contract or some other legal wrong. The architect cannot deal with such claims, unless the employer expressly authorises the architect to do so, and the contractor agrees. However, the architect may be the cause of such claims, for example, by being late in issuing the contractor with necessary information or instructions, or by failing to carry out duties under the contract and the contractor suffers loss as a result.

The case of *Croudace Ltd* v. *London Borough of Lambeth* (1986) establishes this point. In carrying out his duties under the contract the architect owes a duty to the employer to act fairly (*London Borough of Merton* v. *Stanley*

Hugh Leach Ltd (1985)) for example, when certifying: *Sutcliffe* v. *Thackrah* (1974). If the architect fails to carry out the duties under the contract properly, this is a breach of contract for which the employer may be liable in damages. If that happened, no doubt the employer would seek to recover from the architect.

Although the architect owes a duty to the employer to carry out functions under the contract in a professional manner with reasonable care and skill, it is not finally established whether the architect owes a duty of care in tort direct to the contractor. *Pacific Associates Inc* v. *Baxter* (1988), where an earlier, (and some people think) more realistic decision of a very experienced official referee to the contrary was doubted, apparently decided otherwise. In the earlier case (*Michael Sallis & Co Ltd* v. *Calil and William F. Newman & Associates* (1987)) it had been held that an architect owed a duty of care to the contractor to act fairly as between the contractor and the employer in matters such as the issue of certificates and the grant of extensions of time and that the contractor might recover damages direct from an unfair architect. More recent decisions, however, may be signalling a new era of architects' liability to the contractor if reliance can be established: *Henderson* v. *Merritt Syndicates* (1994); *J. Jarvis & Sons Ltd* v. *Castle Wharf Developments and Others* (2001). (See section 4.1.)

10.2 *Extension of time*

10.2.1 Legal principles

At common law, the contractor is bound to complete the work by the date for completion stated in the contract, unless it is prevented from doing so by the employer's fault or breaches of contract (*Percy Bilton Ltd* v. *Greater London Council* (1982)) and the employer's liability extends to the architect's wrongful acts or defaults within the scope of authority. In the absence of an extension of time clause, neither the employer nor the architect would have any power to extend the contract period. Clause 2.7 (clause 2.8 in MWD) deals with extension of the contract period and is linked with clause 2.8 (clause 2.9 in MWD) which provides for liquidated damages. It cannot be over-emphasised that the only effect of the architect granting an extension of time is to fix a new date for completion and, incidentally, to relieve the contractor from paying liquidated damages at the stated rate for the period in question. It certainly does not automatically entitle the contractor to payment of additional preliminaries, as many contractors, architects and even quantity surveyors believe.

10.2.2 Liquidated damages

There are many misconceptions about liquidated damages. The most important point is that the sum stated as such is recoverable whether or not the employer can prove that any loss has been incurred as a result of late completion or even if, in the event, no loss at all has been suffered. On the other hand, the employer cannot recover more than the amount of liquidated damages even if, in the event, the employer actually suffers a much greater loss. The object of the liquidated damages clause is to fix an amount which is a pre-estimate of the quantum of damages which the employer may suffer through late completion and if the contractor is late in completing; it is irrelevant to consider whether in fact there is any loss: *BFI Group of Companies Ltd* v. *DCB Integrated Systems Ltd* (1987), a case on MW 80 holding that liquidated damages are payable even if there is no loss. *Impresa Castelli SpA* v. *Cola Holdings Ltd* (2002) is a later case to much the same effect.

Although often referred to by contractors as 'the penalty clause', a penalty is not enforceable. A sum is treated as liquidated damages (and so recoverable) if it is a fixed and agreed sum which is no more than reasonable and is a genuine pre-estimate of the loss likely to be incurred, or a lesser sum, estimated at the time the contract is made. It matters not that the estimate is a poor one in fact. Deciding whether a sum is liquidated damages or a penalty can be difficult where the parties have inserted a complex set of provisions. However, the courts have indicated that, in deciding the issue, they will not take account of hypothetical situations. They are much more likely to take a pragmatic approach: *Phillips Hong Kong Ltd* v. *The Attorney General of Hong Kong* (1993). It is acceptable to express the liquidated damages as a series of graduated amounts related to the seriousness of the breach: *North Sea Ventilation Ltd* v. *Consafe Engineering (UK) Ltd* (2004).

Under MW and MWD the amount inserted as liquidated damages is usually relatively small in relation to the potential loss to the employer from late completion, but it is bad practice to pluck a figure out of the air. The employer, in consultation with the architect, should have calculated the amount of liquidated damages carefully at pre-tender stage.

To set the figure at the right level, the employer should discuss it carefully with the architect before making the calculation. This is sometimes difficult. In the case of profit-earning assets, there is no problem. All that then need be done is to analyse the likely losses and additional costs. The following should be considered:

- Loss of profit on a new building, e.g. rental income, retail profit, etc.
- Additional supervision and administrative costs – including additional professional fees
- Any other financial results of the late completion, e.g. storage charges for furniture in the case of a domestic building.

An alternative method, but one more open to criticism, is to use one or other of the several formulae used in the public sector, such as that put forward by the Society of Chief Quantity Surveyors in Local Government, which gives an approximation to a detailed analysis of all individual costs, but this calculation should also be based on verifiable data. The figure thus arrived at must be inserted in the contract particulars. If no figure was inserted, no liquidated damages would be payable. The same result would follow if no completion date was inserted since there must be a date from which liquidated damages can run: *Kemp* v. *Rose* (1858). However, liquidated damages are exhaustive of the employer's remedies for late completion and in *Temloc Ltd* v. *Erill Properties Ltd* (1987) the inclusion of '£NIL' in a JCT 80 appendix entry was held to preclude any claim for damages.

Clause 2.8 (clause 2.9 in MWD) makes it clear that liquidated damages are payable at the specified rate only if the Works are not completed by the original completion date or extended contract completion date. They are payable by the contractor at the stated rate per week for the period between the stated completion date and the date of practical completion as certified by the architect under clause 2.9 (clause 2.10 in MWD). The clause confers an express right on the employer to deduct liquidated damages from monies due to the contractor, e.g. progress payments, or the employer may recover them as a debt.

The certificate of practical completion no longer only signifies the date on which the Works have reached that stage. In addition, it certifies that the contractor has sufficiently complied with the CDM Regulations (where applicable) (see section 9.2.1). So, by the time the certificate is issued, the Works may have been at physical practical completion stage for several weeks. The operative date for the end of liquidated damages, therefore, may be the actual date of completion, because clause 2.8.1 (clause 2.9.1 in MWD) states: 'If the Works are not completed . . . '. If damages are to be deducted, the appropriate notices under clauses 4.6.2 or 4.8.3, as appropriate, must be given. Strangely, an additional notice must be given not later than the issue date of the final certificate if the employer wishes to deduct from the final certified sum.

The employer must be careful not to make any representations to the contractor, either orally or in writing, that liquidated damages will not be deducted. This is sometimes done when no extension of time is due, but the employer does not want to frighten the contractor with the prospect of heavy damages. *London Borough of Lewisham* v. *Shepherd Hill Civil Engineering* (2001) suggests that an employer who does make such representations may be estopped (prevented) from later suffering a change of mind and deciding to recover the liquidated damages after all. This will be particularly the case if the contractor, in reliance on the employer's assurances, has paid out money to subcontractors without any deductions.

10.2.3 Extending the contract period

Clause 2.7 (clause 2.8 in MWD) is short and sweet. This is in marked contrast to the long and complicated provisions in other standard form contracts. However, brevity may have been achieved at a price (see below). Clause 2.7 (clause 2.8 in MWD) empowers the architect to grant an extension of time for completion if it becomes apparent that the Works will not be completed *for reasons beyond the control of the contractor*, including compliance with any instruction of the architect the issue of which is not due to a default of the contractor.

It is not clear whether the italicised phrase in fact extends to delay which is the fault of the employer. Sensibly, it ought to do so, but the case law suggests otherwise and this is reinforced by the express reference to architect's instructions. In *Wells* v. *Army & Navy Co-operative Society Ltd* (1902), a very similar phrase, '*other causes beyond the contractor's control*' was held not to extend to delays caused by the employer or his architect. The last sentence of clause 2.7 (clause 2.8 in MWD) was added in the wake of *Scott Lithgow* v. *Secretary of State for Defence* (1989), when the court surprisingly held that, in some circumstances, subcontractors may not be under the control of the contractor. So far as MW and MWD are concerned, it is now clear that subcontractors and suppliers are within the control of the contractor.

In *Peak Construction (Liverpool) Ltd* v. *McKinney Foundations Ltd* (1970), the Court of Appeal ruled that liquidated damages clauses and extension of time clauses were both to be interpreted strictly *contra proferentem* against the employer and should not be read to mean that an employer can recover damages for delay for which the employer was partly to blame. It was said that:

> 'if the employer is in any way responsible for the failure to achieve the completion date, he can recover no liquidated damages at all and is left to prove such general damages as he may have suffered.'

The effect of this is that, although the architect must protect the employer's right to deduct liquidated damages by making extensions of time if appropriate, the courts will interpret that strictly and the architect has no power to make extensions of time for any reasons which do not fall within the grounds set out in the extension of time clause.

That case concerned an in-house form of contract, but its authority was upheld by the Court of Appeal in *Rapid Building Co Ltd* v. *Ealing Family Housing Association Ltd* (1985), a case which involved a JCT 63 contract.

Figure 10.1 illustrates the contractor's duties in claiming an extension of time under clauses 2.7 or 2.8 in MW and MWD respectively.

Figure 10.2 sets out the architect's duties in relation to such a claim. The procedure under clause 2.7 (clause 2.8 of MWD) is straightforward: the contractor must notify the architect in writing if it becomes apparent that the current completion date will not be met. The contract refers to the fact that the Works will not be completed by the date for completion inserted in the contract particulars or any later date fixed. There is no reference to delays to progress.

The architect must then make, in writing, a *reasonable* extension of time. The contract does not say when the architect must do this, but I suggest that the extension of time must be made as soon as possible. Certainly the application should not be put on one side and must be made before the current date for completion is passed, if practicable. The architect must of course be satisfied that the completion date will not be met because of 'reasons beyond the control of the contractor'. The case of *Balfour Beatty Construction Ltd* v. *London Borough of Lambeth* (2002) contains useful guidance for the architect in arriving at the proper extension of time. It is not sufficient for the architect to form an impression of the period; the period must be calculated. It is essential that the architect establishes that the Works will not be completed by the date for completion in the contract or any later extended date before an extension of time is given: *Royal Brompton Hospital* v. *Frederick Alexander Hammond* (2000). Because the contractor has suffered a delay amounting to three weeks does not automatically entitle it to three weeks' extension of time. It is the effect of the three weeks on the completion date that is the crucial factor.

Figure 10.1
Flowchart of contractor's duties in claiming an extension of time

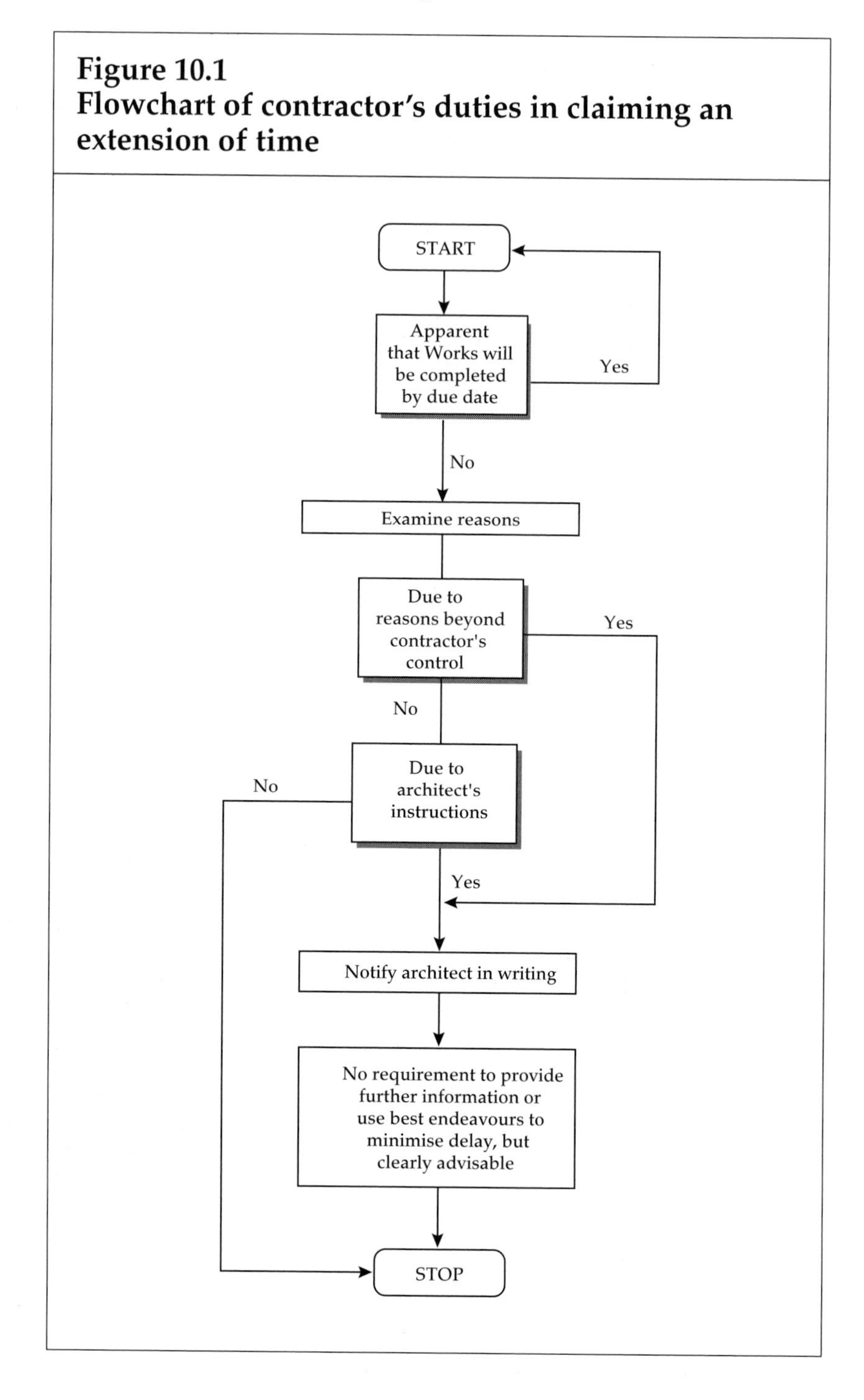

Figure 10.2
Flowchart of architect's duties in relation to claim for extension of time

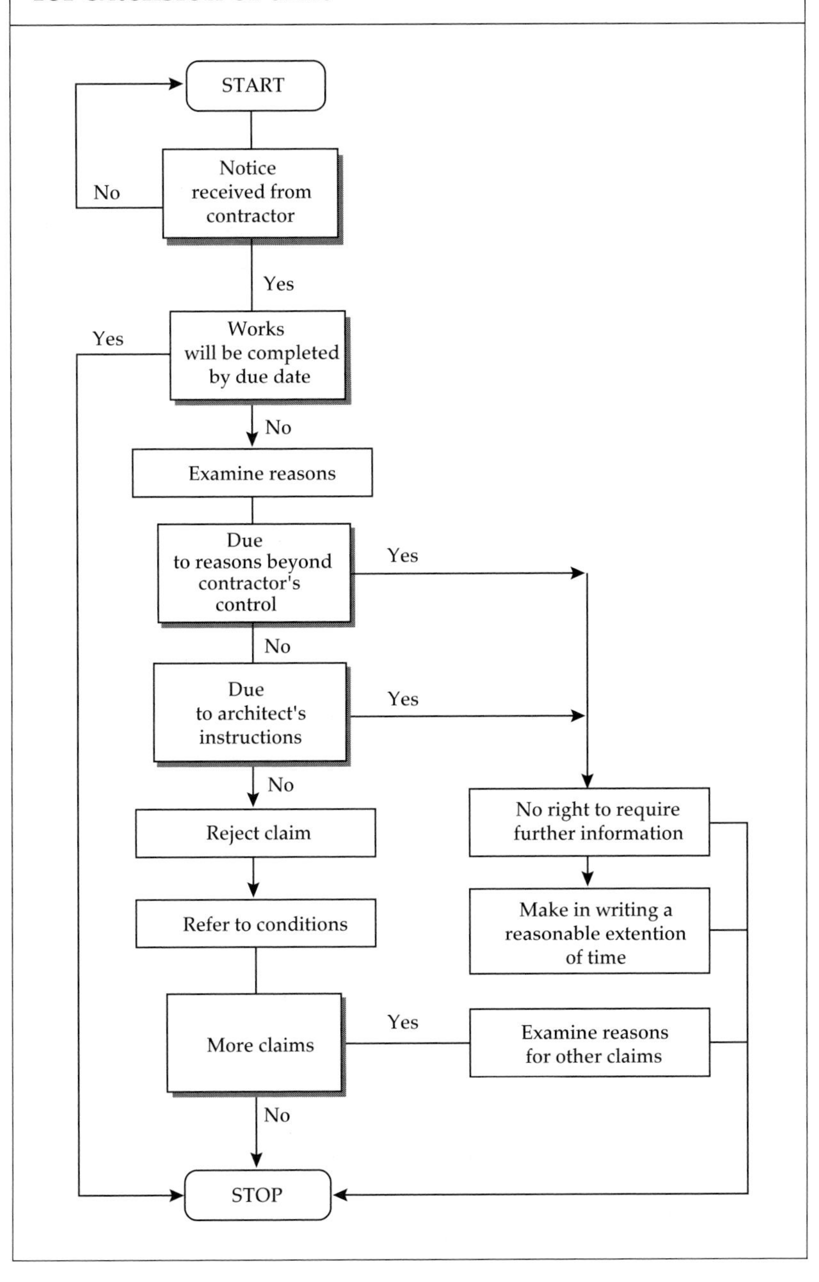

Failure properly to grant an extension of time may result in the contract completion date becoming at large, and liquidated damages becoming unenforceable.

Figure 10.3 is a suggested letter awarding an extension.

What happens if the contractor fails to notify the architect that the completion date will not be met? Clearly the contractor is in breach of contract in not doing so because the obligation rests on the contractor, and it is suggested that the contractor's failure to give notice is a matter which may be taken into account by the architect in determining the extension of time: *London Borough of Merton* v. *Stanley Hugh Leach Ltd* (1985). However, it should be noted that, unlike the position under SBC, the contractor is only obliged to notify those delays which would in principle entitle it to an extension of time. In taking account of the failure, the question to be asked is whether the contractor's failure prejudiced the employer in any way. In other words, if the architect had been informed 'thereupon', could any measures have been taken to reduce or eliminate the delay? The ordinary meaning of 'thereupon' is 'soon' or 'immediately after'. The critical date is the date it became apparent that the Works would not be completed on time. It is certain that the contractor's notice is not a pre-condition ('condition precedent') to the grant of an extension of time and it is up to the architect to monitor progress and in an appropriate case award an extension of time even if the contractor has not given any notification. The primary purpose of an extension of time clause is to protect the employer against the loss of liquidated damages following some act of hindrance or prevention.

10.3 Money claims

10.3.1 General

Although there is limited scope in clause 3.6.3 for the architect to include direct loss and expense in a valuation if it is incurred by the contractor as a direct result of regular progress being affected by compliance with a variation instruction or compliance or noncompliance by the employer with duties in respect of the planning supervisor or principal contractor (if other than the main contractor), there is no clause in the contract which entitles the contractor to make financial claims against the employer whether arising from prolongation or disruption. This leads some people to suppose that this is a risk which the contractor must price. Nothing could be further from the truth. The absence of a direct

Figure 10.3
Letter from architect to contractor making an extension of time under clause 2.7 (clause 2.8 in MWD)

Dear Sirs

PROJECT TITLE

On [*insert date*] you notified me that the works would not be completed by [*insert date for completion or extended date*] because of [*specify reasons, e.g. the national strike of building trades' employees in pursuit of a pay claim*].

I accept that this falls within clause 2.7 [*substitute '2.8' when using MWD*] of the contract, and in accordance therewith I hereby make an extension of time of [*specify period*]. The revised date for completion is now [*insert date*].

Yours faithfully

loss and/or expense clause merely means:

- There is no contractual provision for ascertaining and paying money claims
- The architect has no power to quantify or agree such claims.

The contractor must pursue its claims in arbitration or litigation, unless it and the employer can agree the amount. The contractor is only entitled to claim in this way for what can be termed 'common law claims', usually breaches of contract. The absence of a claims clause is not necessarily a benefit as some employers appear to think. The situation can, of course, be overcome by including a suitably worded clause as an amendment to the contract. There are plenty of precedents, for example, IC, clauses 4.17 and 4.18.

10.3.2 Types of claims

Financial claims are commonly referred to in the industry as *ex-contractual claims*. These are claims made outside the express provisions of the contract, usually as a result of breach of its terms, express or implied. They are also called common law claims, and increasingly are based on *implied terms* relating to co-operation and non-interference by the employer.

The situation is well-illustrated by *Holland Hannen & Cubitt (Northern) Ltd* v. *Welsh Health Technical Services Organisation* (1981) where it was pleaded on behalf of contractors under a JCT 63 contract that there were implied terms whereby the employer contracted that the employer and the architect:

- Would do all things necessary on their part to enable [the contractors] to carry out and complete the works expeditiously, economically and in accordance with the contract.

Conversely, that neither the employer nor the architect:

- Would in any way hinder or prevent [the contractors] from carrying out and completing the works expeditiously, economically and in accordance with the contract.

It is suggested that such terms are to be implied when the contract is in MW or MWD form and this being so opens up a wide area for claims, which may also arise in tort.

Table 10.1
Clauses that may give rise to claims under MW and MWD

Clause	Event	Type
2.2 (2.3 under MWD)	Failure to allow commencement on the due date by lack of possession or otherwise	CL
2.3 (2.4 under MWD)	Failure to issue further necessary information	CL
2.4 (2.5 under MWD)	Errors or inconsistencies in the contract documents	C
2.5 (2.6 under MWD)	Divergence between statutory requirements and contract documents or architect's instruction	C
2.7 (2.8 under MWD)	Failure to give extension adequately or in good time	CL
2.8 (2.9 under MWD)	Wrongful deduction of damages	CL
2.9 (2.10 under MWD)	Failure to certify practical completion at the proper time	CL
2.10 (2.11 under MWD)	Wrongful inclusion of work not being defects etc.	CL
2.11 (2.12 under MWD)	Failure to issue certificate that the contractor has discharged its obligations	CL
3.1	Assignment without consent	CL
3.3.1	Unreasonably withholding consent to subletting	CL
3.4	Failure to confirm oral instructions Instructions altering the whole character or scope of the work Wrongful employment of others to do work	CL CL CL
3.6	Variations Compliance or non-compliance of the employer with clause 3.9 Wrongful omission of work to be done by others	C C CL
3.8	Unreasonably or vexatiously instructing removal of employees from Works	CL

Table 10.1 *Contd*

Clause	Event	Type
4.3	Failure to certify payment on the due dates Failure to certify payment for materials properly on site	CL CL
4.5	Failure to certify payment at practical completion	CL
4.8.1	Failure to issue final certificate	CL
4.11	Contribution, levy and tax changes	C
6.4	Invalid termination Payment after termination	CL C
6.5	Invalid termination	CL
6.11.4	Payment after termination	C

Key
C = Contractual claims CL = Common law claims
Contractual claims are usually dealt with by the architect.
Common law claims are usually dealt with by the employer.
Note: There is no loss and/or expense clause in this contract. Such claims have to be made at common law.

If claims of this nature are made by the contractor they must be made against the employer. The employer may well seek the architect's advice, and possibly the architect may be authorised to deal with them. But the contractor is not entitled to reimbursement because it is losing money; it is up to the contractor to establish that the employer or the architect is in breach. Valid claims can only arise because the contractor suffers loss through the fault of the employer or those for whom the employer is responsible in law.

Table 10.1 summarises the MW and MWD clauses which may give rise to claims. Useful books for reference purposes are:

- *Building Contract Claims*, 4th edition, by David Chappell, Vincent Powell-Smith and John Sims, Blackwell Publishing, 2005.
- *Causation and Delay in Construction Disputes*, 2nd edition, by Nicholas J. Carnell, Blackwell Publishing, 2005.

10.4 *Summary*

Claims for time and money are distinct; there is no necessary connection between the two.

Liquidated damages

Liquidated damages are:

- A genuine pre-estimate of likely loss or a lesser sum
- Recoverable without proof of loss
- Recoverable only if the contractor has not completed the Works by the original or extended completion date.

Extension of time

The architect is bound to make a reasonable extension of time for completion if:

- The contractor gives written notice that the Works will not be completed by the current completion date because of reasons beyond its control, including an architect's instruction not occasioned by the contractor's default.

If the architect fails to grant an extension of time for completion or fails to grant it before the date for completion, the contract time may become 'at large' and liquidated damages will be irrecoverable.

Money claims

- There is limited contractual provision for the contractor to be reimbursed for direct loss and/or expense
- Any claims by the contractor must be pursued under the general law
- Such claims can only arise because the contractor suffers loss through the architect's fault or that of the employer.

CHAPTER ELEVEN
PAYMENT

11.1 Contract sum

The contract sum is the sum of money which is inserted in Article 2 of the Articles of Agreement. It is stated to be exclusive of VAT which means that VAT payments which may be necessary will be additional to this sum (clause 4.1). How far this will affect the employer will depend on the employer's status, from the point of view of being able to reclaim VAT, and the work involved in the contract.

The figure in Article 2 will be the contractor's tender figure or such figure as the parties agree, perhaps after negotiation. The importance of the sum cannot be over-emphasised. It is the sum for which the contractor has agreed to carry out the whole of the Works as shown on the contract documents. MW and MWD are lump sum contracts which means that the contractor is entitled to payment provided it completes substantially the whole of the Works. In theory, the existence of a system of interim payments does not alter the position and if the contractor abandons the work before completion, the employer is entitled to pay nothing more. Once written into the contract, the contract sum may be adjusted only in accordance with the contract provisions (Table 11.1). Errors or omissions in the computation of the contract sum are deemed to be accepted by employer and contractor. Inconsistencies may be corrected in accordance with clause 2.4 (clause 2.5 in MWD) (see section 3.1.2). It is quite possible for the contractor to make a considerable error in its calculations to such an extent that the contract becomes no longer viable from its point of view. If the error is undetected before the contract is entered into, there is nothing the contractor can do about the situation, except perhaps submit an *ex-gratia* (on grounds of hardship) claim, with little hope of success. This situation must be avoided if possible because it is unsatisfactory from all points of view. The employer may indeed think that there is a financial advantage to be gained, but a contractor in this position has very little incentive to work efficiently and every reason to submit claims for additional payment at every opportunity.

Table 11.1
Adjustment of contract sum under MW and MWD

Clause	Adjustment
2.4 (2.5 under MWD)	Inconsistencies in the contract document
2.10 (2.11 under MWD)	Defects etc. during the rectification period
3.5	Non-compliance with instructions
3.6	Variations
3.7	Provisional sums
4.8.1	Computation of the final amount
4.11	Contribution, levy and tax changes
5.4A	Insurance money
5.4B	Making good of loss or damage

Unfortunately, unless the contract documentation includes quantified schedules or bills of quantities, it is very difficult to check the contractor's pricing. Some errors may be obvious, but where the contract documents consist of drawings and specification, the contractor's pricing strategy may be obscure unless it separately submits a detailed breakdown of the figure.

11.2 *Payment before practical completion*

Clause 4.3 sets out the procedure for what is termed 'progress payments'. The architect must certify progress payments at four-weekly intervals. The certification does not depend upon the contractor's request. The certification dates can be pinpointed, because they are calculated from the date of commencement stated in the contract particulars. In order to give the contract business efficacy, the architect must issue certificates to the employer, such certificates being certificates of progress payments to the contractor. This view is reinforced by reference, later in the clause, to the employer paying 'the amount so certified'. Clearly, the employer cannot pay these amounts without knowing what they are. The employer must receive the certificate, while the contractor receives a copy. The clause states that the certificate must state to what the payment relates and the basis on which the amount was calculated. This is anything but clear. It may mean that the architect must include, as part of the certificate, a complete breakdown of the valuation showing each element of work and materials and the value of each element. Alternatively, it may be sufficient for the architect to identify it as certifying work up to a specified date and based on the prices in the priced specification or schedule of rates. On balance and in view of the absence of any independent valuation by a quantity surveyor, it is considered that the first option (a full valuation) is required. However, the matter is not free from doubt.

The employer must pay within 14 days of the date of issue of the certificate. In practice, it is usual for the contractor to send four-weekly statements of account for the architect's consideration before valuation and certification.

Although there is no provision for any other system of payment, it may be more convenient, on small works, to agree a system of stage payments. If it is desired to operate in this way, it is important to make the necessary amendments to the printed conditions and to ensure that the contractor is aware of the change at tender stage. It will have a considerable impact on its pricing strategy.

The amount to be included in the certificate is to consist of:

- The percentage stated in the contract particulars (usually 95%) of the total value of work properly executed
- Amounts ascertained or agreed under clause 3.6 (variations) or clause 3.7 (provisional sums)
- The percentage stated in the contract particulars (usually 95%) of the value of any materials and goods which have been reasonably and properly brought on the site for the purpose of the Works and which are adequately stored and protected against the weather and other casualties
- Amounts calculated under the provisions of clause 4.11.

Less

- The total amount due in previous certificates.

Each of the above items requires careful consideration, as follows.

The percentage of the total value of work properly executed

There has been a difference of opinion about the meaning of the word 'value' in this context. The generally accepted view is that it is the valuation obtained by means of reference to the priced document in relation to the amount of work actually carried out. Defective work is not included and a retention percentage is deducted. The retention is intended to deal with problems which might arise (see section 11.7). The biggest problem would be if the contractor went into liquidation immediately following a payment. The alternative view of the meaning of 'value' derives from this possibility. From the employer's point of view, the value of the contractor's work is the value of the whole contract less the cost of completing the work with the aid of another contractor and additional professional fees. The additional cost could be considerable and incapable of being met from the retention fund.

The latter view finds little favour with contractors since the certificates issued at any given stage would bear no relation to the money expended by the contractor. Indeed, during the early part of a job, the certificates might even show a minus figure. The two views might be termed contractor value and employer value. The services of a quantity surveyor would be necessary to determine the probable cost of completion at the time of each certification. The task would not be easy. It is possible to operate this system if it is made clear to the contractor at the time of tender so that it can take extra financing charges into account.

Case law suggests that the courts will favour the traditional contractor value view of 'value'.

'Properly executed' refers to the fact that the architect must not include the value of work which is defective, that is, not in accordance with the contract. If the architect does certify defective work because, perhaps, the defect does not make itself immediately apparent, the correct procedure is to omit the value from the next certificate. This should pose no difficulties unless, of course, the contractor abandons the work first.

Amounts ascertained or agreed under clauses 3.6 and 3.7

This refers to any variations which are valued before the date of the certificate and any instructions which the architect may issue with regard to provisional sums which result in an adjustment to be valued under clause 3.6.

The value of materials and goods, etc.

The reasoning behind the inclusion of payments for unfixed materials on site is clear. It is to enable the contractor to recover, at the earliest possible time, money which it has already laid out. The architect need not include any materials which are considered to have been delivered to the site unreasonably early for the sole purpose of obtaining payment, but in its present form the clause contains serious dangers for the employer. There is no provision for the contractor to provide proof of ownership before payment. Therefore, if unfixed materials are included in a certificate and the contractor does not own them, the employer could be faced with the prospect of paying twice for the same materials if the contractor goes into liquidation and the true owner claims the goods from site: *Dawber Williamson Roofing Co Ltd* v. *Humberside County Council* (1979). The prevalence of retention of title clauses in the supply contracts of builders' merchants makes this a very real danger. Such a retention of title clause cannot be overcome by anything which may be written into the contract. Many architects amend the clause by deleting the whole of the provision in clause 4.3.2. Since there is no provision for payment for off-site materials, this will then mean that the architect does not have to certify any unfixed materials whatsoever.

The total amount due in previous certificates

Previously this clause referred to previous payments made by the employer – something of which the architect would have no formal knowledge.

To comply with the Housing Grants, Construction and Regeneration Act 1996, the employer must give the contractor a written notice, not later than five days after the issue of a certificate. The written notice must specify the amount of payment to be made. It probably does not matter if the employer forgets to give the notice provided that the amount certified is paid, and paid within the stipulated 14 days (see clause 4.6.1).

11.3 *Penultimate certificate*

The contract provides for a special payment to be made at practical completion (clause 4.5). This clause is entitled 'Penultimate certificate' because, although it is not expressly stated, it is clear that regular certificates will cease at this point because there is no more work to value. Indeed the only other payment to be made will be the final payment.

The penultimate certificate must be issued within 14 days after the date of practical completion as certified by the architect. The employer has 14 days, as before, in which to pay. The effect of the certificate is to release to the contractor the balance of the monies due on the full value of the Works, plus half the retention already held. This usually means that the employer retains only $2\frac{1}{2}\%$ of the total value. Obviously, it may not be possible for the architect to know the full value precisely at this stage, but it is to be certified so far as that amount is ascertainable at the date of practical completion. In other words, the architect must use reasonable endeavours to ascertain the amount pending final computation. Amounts ascertained or agreed under clauses 3.6 and 3.7 are to be included and, again, reference is made to deducting the amounts of previous progress payments. The comments made earlier are also applicable to this clause, including the written notice of payment to be given by the employer.

11.4 *Final certificate*

The contract lays down a precise time sequence for the events leading up to, and the issue of, the final certificate (see the contract time chart, Figure 11.1).

The contractor's duty is to send the architect all the documentation the architect reasonably requires in order to compute the amount to be finally certified. The architect is probably entitled to request any particular supporting evidence necessary. The contractor has three months, or such other period as is inserted in the contract particulars, from the date of practical completion to send the information to the architect.

Figure 11.1
MW and MWD time chart

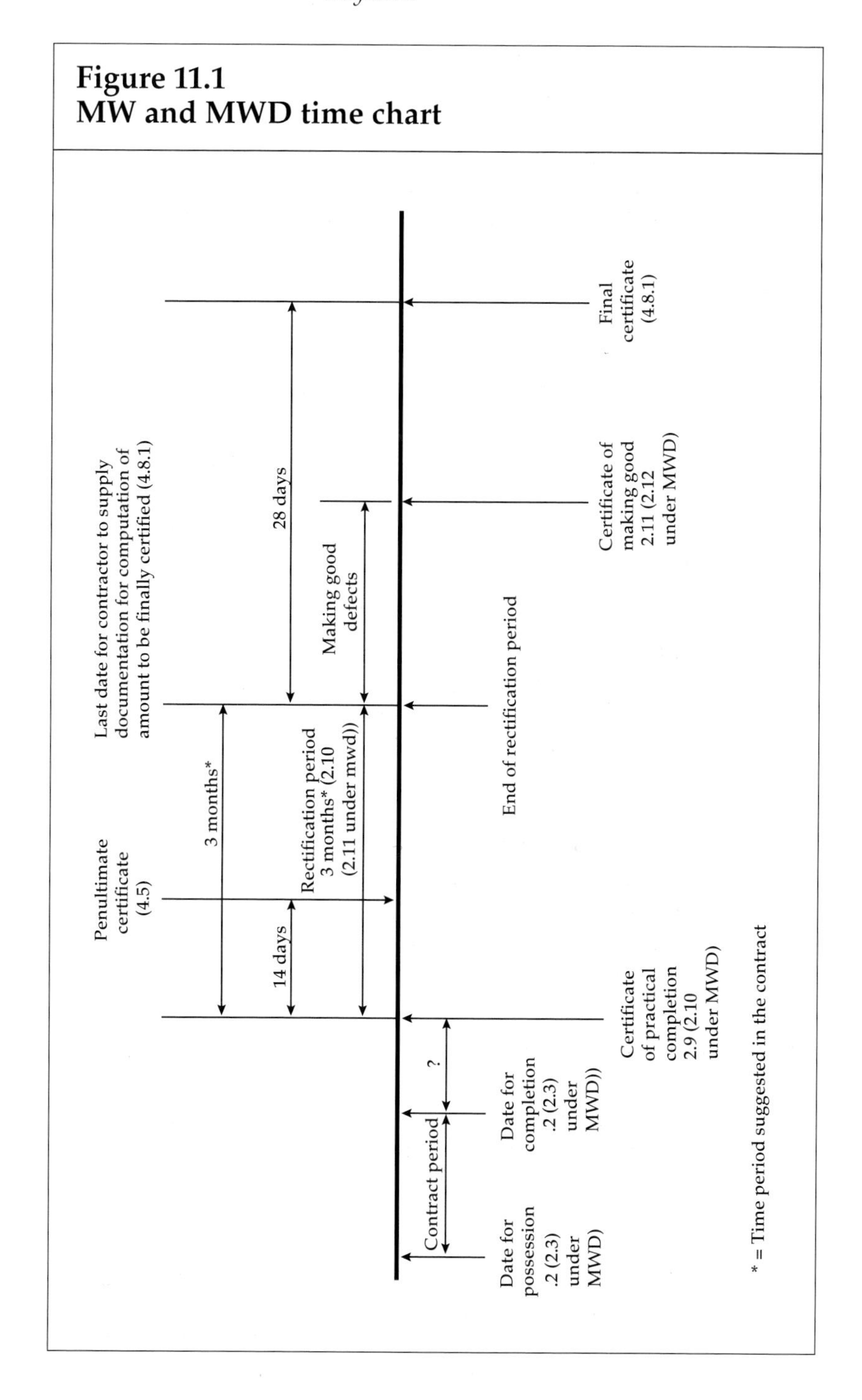

Although not expressly stated, it is clear that this three-month period is relevant to the three-month period in clause 2.10 (clause 2.11 in MWD), so that if the rectification period referred to in clause 2.10 (2.11) is increased to, say, 12 months, the period referred to in clause 4.8.1 must be the same. This is because clause 2.10 (2.11) gives the architect power to issue instructions about the making good of defects. It may be that it involves some adjustment to the contract sum, in which case the amount must be reflected in documents to be provided by the contractor under clause 4.8.1 In such circumstances, different time periods are not workable. The architect must issue the final certificate within 28 days of the receipt of the contractor's information. There is a proviso. The architect may not issue the final certificate until after the certificate of making good under clause 2.11 (clause 2.12 in MWD) has been issued. In practice, this will often be the deciding factor. The certificate of making good must precede the final certificate. That is clear from the use of two different tenses and the proviso: 'provided the Architect... *has issued* the certificate under clause 2.11' [clause 2.12 in MWD] and '*issue* a final certificate' (emphasis added). The safest way is for the architect to issue the certificate of making good one day and the final certificate on the next day. In theory one certificate could be issued just a few minutes before the other. Obviously, they cannot be issued together. If the contractor is late in sending the information to the architect, the contractor is technically in breach, but it is of little consequence. The contractor is simply delaying the time when it receives payment because the architect's 28 days does not begin to run until the contractor's documentation is received.

However, failure by the contractor to supply the information required under clause 4.8.1 ought not to delay the issue of the final certificate. Throughout the contract period, the architect should have been keeping a running total of the prospective final account and obtaining documents from the contractor where necessary to carry out the valuations under clause 3.6. The period after practical completion is to allow the contractor to draw the architect's attention to any further information that is relevant. It is not intended that the contractor submits nothing until this point and then submits vast bundles of documents. If that is what had been intended, 28 days would have been an entirely inadequate period for the architect to prepare the final account. Once the certificate of making good has been issued, it is suggested that lack of information from the contractor must not delay the issue of the final certificate. Although it is customary for the contractor to submit its version of the final account during this period, it is clear from the contract that it is the architect who must compute the final account.

The final certificate must state the amount remaining due to the contractor or, more unusually, the amount due to the employer. The latter situation will only occur if the architect has previously overcertified. The certificate should state to what the amount relates and how it has been calculated. Although not expressly required under the contract, the architect should have already provided the contractor with a final account and reference to this should be sufficient. The employer has 14 days, as before, in which to pay from the date of the final certificate. Clause 4.8.2 stipulates that the employer must give written notice to the contractor specifying the amount of payment to be made. The notice must be given not later than five days after the issue date of the final certificate. This is a similar provision to the notice provision for interim certificates and it is for the same purpose.

There is no provision in the contract for the contractor to agree the computations before the final certificate is issued, but it is customary to attempt to obtain the contractor's agreement. If the contractor delays sending agreement to the architect this does not affect the architect's duty to issue the final certificate within 28 days: *Penwith District Council* v. *V. P. Developments Ltd* (1999). A letter from the architect to the contractor should make this clear (Figure 11.2).

11.5 *Effect of certificate*

It is refreshing to note that no certificate is stated to be conclusive. Thus, the issue of the certificate of making good does not preclude the employer from asserting that the contractor is liable for anything which is not in accordance with the contract. Similarly, if the architect inadvertently includes defective work in a four-weekly progress certificate, the situation can be remedied with the next certificate issued.

The contractor's liability is not reduced in any way by the issue of the final certificate which is not even conclusive as far as the computations are concerned. *Crown Estates Commissioners* v. *John Mowlem* (1994) considerably broadened the conclusive effect of the final certificate issued under JCT 80. It is considered that this case has no relevance to MW or MWD.

11.6 *Interest and withholding payment*

If the employer fails to pay any amount due to the contractor by the final date for payment (i.e. 14 days from date of issue of a certificate), clause 4.4

Figure 11.2
Letter from architect to contractor requesting agreement to the computation of the final sum

Dear Sir

PROJECT TITLE

I enclose copies of the computation of the final sum to be certified.

I should be pleased if you would signify your agreement to the sum and the way in which it has been calculated by signing and dating one copy of the calculation in the space provided and returning it to me by the [*insert date*] at the latest.

If you have any queries, please telephone me as soon as possible, but you should note that, in any event, I have a duty under clause 4.8.1 of the conditions of contract to issue my final certificate no later than [*insert date*].

Yours faithfully

Copy: Employer

in respect of interim and the penultimate certificates, and clause 4.9 in respect of the final certificate, require the employer to pay simple interest on the outstanding amount at the rate of 5% over the dealing rate of the Bank of England current on the date payment becomes overdue. This is a substantial amount although not as substantial as the current percentage under the Late Payment of Commercial Debts (Interest) Act 1998 which requires 8% above Base Rate together with a lump sum payment.

The employer may not withhold or deduct any amount from a payment to the contractor, without first giving a written notice to the contractor not later than five days before the final date for payment. The notice must state the grounds for withholding payment and the amount of money withheld in respect of each ground. The information must be detailed enough to enable the contractor to understand the reason why it is not receiving the amount withheld. The contractor may, of course, seek immediate adjudication under the provisions of Article 6 (see Chapter 14). This is something about which the architect should advise the employer in good time.

11.7 Retention

The reference to retention in this contract is remarkably brief. It is mentioned only in clauses 4.3 and 4.5 and then only in respect of the amount. In particular there is no reference to:

- The purposes for which the retention can be used
- Its status as trust money
- Keeping it in a separate bank account.

The reason for this is clearly to maintain the brevity of the contract as a whole, but there may be repercussions:

Use of retention

Clause 3.5 used to allow the employer to deduct the cost of employing others to carry out instructions with which the contractor had failed to comply, from monies due to the contractor. Retention monies clearly fell within this clause, but now the cost is to be deducted from the contract sum. If the contractor goes into liquidation before the Works are complete, there is no doubt that the employer would be entitled to make use of retention monies to complete the work. There is no express provision enabling the employer to make use of the retention for any other purpose. For example, the employer may not deduct the cost of taking out insurance if the contractor defaults. In practice, as mentioned earlier, the situation may suggest taking a broader view.

Trust money

Unlike the corresponding provisions of SBC, IC and ICD, the retention is not expressly stated to be trust money and it is not considered that any term would be implied from the general law to create such a trust.

Whether the retention is or is not a trust is of vital importance to the contractor for two reasons. First, a trust is governed by statute and, despite what the contract provisions may state, it is possible that the employer always has a duty to invest and to return any interest to the contractor. Where the retention is not stated to be a trust, the contractor is not entitled to any interest.

Second, where the retention is stated to be trust money, the employer is in the position of trustee, merely holding the money for the contractor's ultimate benefit. It is, in no sense, the employer's money. Therefore, if the employer were to become insolvent, the trust money would not form part of the employer's assets in the hands of the trustee in bankruptcy or the liquidator. The contractor would be able to recover the full amount provided that it was clearly separated from the employer's other money and held in a separate account. If the money is not a trust, as in the case of MW and MWD, the contractor would have no better chance of recovery than any other unsecured creditor. MW and MWD, therefore, leave the contractor at a severe disadvantage.

Separate bank account

Where the retention is trust money, it is often required to be placed by the employer in a separate bank account specially opened for the purpose. The result is that, in the event of the employer becoming insolvent, there is no difficulty in identifying the money. It is also good practice for any trust money to be dealt with separately from the employer's own money so that the investment income can be easily ascertained. Even if the contract is silent, it is now established law that the contractor can demand that trust money be kept in a separate bank account. The employer has a duty to put trust monies in a separate bank account whether or not so requested. Certainly, the employer may not mix trust money with working capital: *Wates Construction (London) Ltd* v. *Franthom Property Ltd* (1991). In the present case, since the retention is not stated to be trust money, the contractor cannot so demand. However, it can be made into trust money by a suitable amendment to the contract if that is required.

11.8 Variations

Variations are covered by clause 3.6. It gives the architect power to order:

- In the case of MWD, a change in the Employer's Requirements resulting in a change to the design of the CDP work
- An addition to the Works
- An omission from the Works
- A change in the Works
- A change in the order in which they are to be carried out
- A change in the period in which they are to be carried out.

It should be noted that the contractor has no right to object to any change in sequence of the Works.

This seems rather harsh, but taken in the context of the size and type of contract likely to be carried out under this form it is unusual to find problems arising in practice. In any case, the contractor is entitled to be paid for changing the sequence or timing, so, in theory, it should not suffer. The clause provides that the architect is to value all the types of variation listed. The wise architect will leave such matters to the quantity surveyor, if employed, but the contract is silent about the role of the quantity surveyor (see section 7.1). Even if the architect does ask the quantity surveyor to undertake the valuation of variations, the architect bears the ultimate responsibility.

The contract provides in clause 3.6.2 that the architect must endeavour to agree a price with the contractor for any variation, but agreement must be reached before the contractor carries out the instruction. In practice, it allows the architect to invite and accept the contractor's quotation. On small works this is the sensible way of operating the provision if the additional or omitted work is anything other than more, or less, of the same.

If the architect does not agree a price with the contractor, clause 3.6 lays down how the valuation is to be carried out. It is to be done on a 'fair and reasonable basis'. Although it is pleasant to think that 'fair and reasonable' is an objective concept, in practice it is the architect's opinion as to what is fair and reasonable that matters. When carrying out the valuation, the architect must use, where relevant, prices in the relevant priced document. It is:

- The priced specification *or*
- The priced work schedules *or*
- The contractor's own schedule of rates.

It is a matter for the architect to decide whether the prices are relevant. It is considered that unless the work being valued is precisely the same, carried out under the same conditions, the architect is probably at liberty to ignore the prices in the priced document. This is because even a slight change in the conditions under which work is being carried out will have a marked effect on the cost to the contractor. The status of the contractor's schedule of rates has already been discussed (Chapter 3). If the architect waits until the first variation has to be valued before asking the contractor to provide it, the way is open for an unscrupulous contractor to massage the rates. Whether or not the contractor does so, the architect will retain suspicions. If the schedule is not provided before the job starts on site, any later provision of the schedule will be suspect and the architect will have an unenviable task trying to value variations properly.

The valuation must include any direct loss and/or expense due to regular progress of the Works being affected by compliance with the variation instruction or the employer's compliance or non-compliance with clause 3.9. The purpose of this provision is to reimburse the contractor for the loss and/or expense directly resulting from the variation, but not forming part of the cost of the varied work itself. In other words, it covers the effect of the introduction of the variation upon other unvaried work together with such things as scaffolding and, if the contract period is prolonged as a result, it also covers the extra site establishment costs. It is suggested that it also covers other items of cost not directly related to the variation, which cannot be covered by prices in the priced documents.

The reference to clause 3.9 is to the employer's very broad duty to comply with the CDM Regulations. Particular duties are highlighted in clause 3.9.1: the duty to ensure that the planning supervisor carries out relevant duties under the CDM Regulations and, if the contractor is not the principal contractor, the employer must also ensure that the principal contractor carries out all relevant duties under the Regulations. It should be noted that it is not just the employer's failure to comply with these duties, but also actual compliance, which entitles the contractor to such loss and/or expense as it may incur.

There is no express requirement for the contractor to submit vouchers or other information to assist the architect in arriving at a fair and reasonable valuation, but it would be a foolish contractor who refused reasonable requests in this respect. If the contractor refuses to supply information, the architect must carry out the valuation using reasonable endeavours and the contractor will have no valid claim if it receives less than it expects. The valuation will, of course, include an element for profit, overheads and so on, as usual.

Clearly a fair and reasonable valuation must also include the valuation of work or conditions not expressly covered in the instruction but affected by that instruction. For example, if the architect issues an instruction to change the doors from painted plywood faced to natural hardwood veneered and varnished, it might well affect the sequence in which the contractor hangs the doors, the degree of protection required and the difficulty of painting surrounding woodwork. All this must be taken into account in the valuation; it is 'direct loss and/or expense' associated with the variation, as is other disruption or prolongation directly flowing from the variation.

11.9 Order of work

Although clause 3.6 refers to the architect's power to change the order or period in which the Works are to be carried out, it is by no means clear what this means. Although the order of the work may be changed if an order is already stated in the specification or schedules, there is no power to create an order. Moreover, there is no provision for the architect to insert any dates against parts of the Works. Therefore, the contractor is under no obligation to complete any parts of the Works before the contract date for completion.

In that context, the reference to 'period' may be irrelevant. It cannot mean that the architect is entitled to vary the date for completion, because that power already exists to a limited extent in clause 2.7 (clause 2.8 in MWD) (extension of time). If it means that the architect can reduce the contract period, it amounts to an acceleration clause. It is thought that an acceleration clause would have to be signalled in a much more positive way: *J. Spurling Ltd* v. *Bradshaw* [1956].

11.10 Provisional sums

Clause 3.7 empowers the architect to issue instructions to the contractor directing how provisional sums are to be expended. Provisional sums are usually included when the precise cost or extent of work is not known at the time of tender. The purpose is to have a sum of money to cover the cost. When the architect issues the instruction, the provisional sum must be omitted and the instruction must be valued in accordance with the principles in clause 3.6 (explained in section 11.8 above). It is possible to use clause 3.7 to nominate a subcontractor, but this is not a sensible course (see section 8.2.3).

11.11 *Fluctuations*

Clauses 4.10 and 4.11 deal with fluctuations. Clause 4.11 is used if it is intended that the bare minimum fluctuations to deal with contribution, levy and tax changes are to be allowed. Detailed provisions are to be found in schedule 2. In the case of short-term projects for which this form is intended to be used, this provision will be deleted in the contract particulars.

Clause 4.10 states that, save for the limited provision of clause 4.11 if retained, the contract is to be considered fixed price. The contractor must carry the risk of increase in cost of labour, materials, plant and other resources. Of course, in the unlikely event of a general fall in costs, the contractor would gain.

11.12 *Summary*

Contract sum

- Exclusive of VAT
- The amount for which the contractor agrees to carry out the whole of the work
- MW and MWD are 'lump sum' contracts
- May be adjusted only in accordance with the contract provisions
- Errors are deemed to be accepted by both parties.

Payment before practical completion

- The architect is to certify payments at not less than four-weekly intervals
- An alternative system of payment may be agreed between the parties
- The employer must pay within 14 days of the date of the certificate
- The employer must give five days' notice of the amount to be paid
- The architect must decide the meaning of 'value' and inform the contractor at tender stage
- The employer may reserve a retention on the whole of the certified sum
- There are dangers in certifying unfixed materials.

Penultimate certificate

- The certificate must be issued within 14 days of the date of practical completion
- The employer must give five days' notice of the amount to be paid
- Half the retention must be released.

Final certificate

- The contractor must send all the documents the architect requires to compute the final sum
- The contractor should have three months or the same as the rectification period from the date of practical completion to send them
- The final certificate must be issued within 28 days of the receipt of the contractor's information provided that the architect has issued the certificate of making good
- The employer must give five days' notice of the amount to be paid
- The contractor does not have to agree the architect's computations.

Effect of certificates

- No certificate is conclusive in any respect
- The contractor's liability is not reduced by the final certificate.

Withholding payment

- Notice must be given five days before the payment is finally due
- 5% simple interest above Bank of England's official dealing rate is payable on amounts not properly paid.

Retention

- The employer is probably entitled to use money from the retention fund if the contractor defaults
- The retention is not trust money and need not be kept in a separate bank account
- The contractor has no better claim on the retention money than any other unsecured creditor if the employer becomes insolvent.

Variations

- The contractor cannot object to a change in the sequence of the Works
- The architect must check all valuations

- The architect may agree a price with the contractor before an instruction is carried out
- Otherwise valuations must be on a fair and reasonable basis using prices in the priced document if relevant
- There is no provision for the valuation of claims for loss and/or expense
- Provisional sum work must be valued in accordance with clause 3.6.

CHAPTER TWELVE
TERMINATION

12.1 *General*

Under the general law, a serious breach by one of the parties to a contract may entitle the other (innocent) party to treat its obligations under the contract as at an end. But it is not every breach of contract which entitles the innocent party to behave in this way. The breach must be 'repudiatory', i.e. it must be conduct which makes it plain that the party in breach will not perform its obligations, or else it must consist of misperformance which goes to the root of the contract.

Until all the facts have been investigated – and this must take place after the event – it is sometimes difficult to say at what point a breach has occurred that is sufficiently serious to entitle the innocent party to accept the breach and bring the obligations of both parties to an end.

In common with most standard building contracts, MW and MWD provide for either party to terminate the contractor's employment under the contract by going through a prescribed procedure. The termination clause (6) attempts to improve on the common law rights of the parties; indeed, the clause goes on to specify what the rights of the parties are after there has been a valid termination. Clause 6.2.3 provides for notices to be given by actual, special or recorded delivery.

Termination of the contractor's employment is a serious step, and the consequences of a wrongful termination are serious. It has been held, however, that where a party honestly relies on a contract provision, although mistaken, the party has not repudiated the contract: *Woodar Investment Development Ltd* v. *Wimpey Construction UK Ltd* (1980). Termination is best avoided if possible, and the process is fraught with pitfalls for the unwary. Quite apart from the possibility of an action for breach of contract by one party or the other if things go wrong, the employer is always placed in an impossible position as far as getting the project completed is concerned.

Even if the employer is successful in recovering costs from the contractor, the time which has been lost can never be recovered. In practice, the contractor is in an invidious position, whether it terminates its own employment or the employer does so. The formalities laid down in the termination provisions must be followed exactly if a costly dispute is to be avoided.

The architect bears a heavy burden in having to advise the employer about rights and the procedure to be followed, and the termination clauses are only too easy to misunderstand. A contractor who believes that it has grounds on which to terminate its own employment under clause 6 should never attempt to do so without seeking competent legal advice.

Either the employer or the contractor may exercise the right to terminate the contractor's employment and MW and MWD deal separately with termination by the employer and by the contractor.

12.2 *Termination by the employer*

12.2.1 Grounds and procedure

The procedure for termination is set out in Figure 12.1 while the grounds which may give rise to termination and the procedure to be followed are specified in clauses 6.4, 6.5 and 6.6.

There are five separate grounds for termination in that clause. They are that the contractor:

(1) Wholly or substantially suspends the carrying out of the Works before practical completion without reasonable cause (clause 6.4.1.1) *or*
(2) Fails to proceed regularly and diligently with the Works before practical completion without reasonable cause (clause 6.4.1.2) *or*
(3) Fails to comply with the CDM Regulations before practical completion (clause 6.4.1.3)
(4) Becomes insolvent in one of the ways specified in clause 6.1
(5) Commits a corrupt act (clause 6.6).

The first three of these grounds are described as a 'default'.

If the employer, on the architect's advice, decides to terminate the contractor's employment, the architect must ensure that the procedure is followed precisely. There is provision for preliminary notice before the right of termination is exercised. The architect must give notice which specifies the default and which requires it to be ended. Figure 12.2 is the sort of letter which the architect might draft and it must be sent by

Figure 12.1
Flowchart of termination by employer

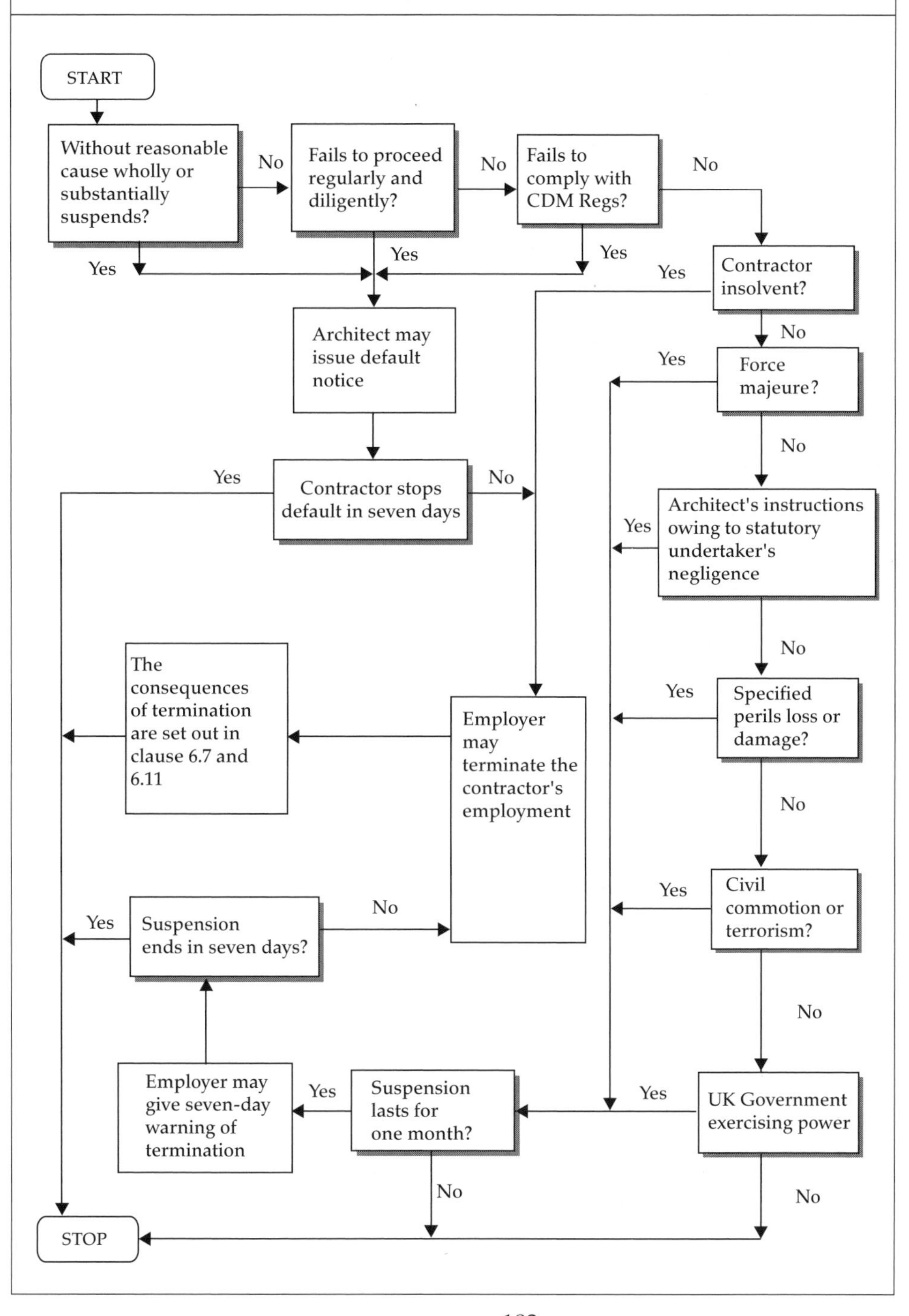

Figure 12.2
Letter from architect to contractor giving notice of default

SPECIAL, RECORDED OR ACTUAL DELIVERY

Dear Sir(s)

PROJECT TITLE

I hereby give notice under clause 6.4.1 of the contract that you are in default in the following respects:

[*Insert details of the default with dates if appropriate*]

If you continue the default for seven days after receipt of this notice, the employer may thereupon terminate your employment under this contract without further notice.

Yours faithfully

Copies: Employer
Quantity surveyor [*if appointed*]

special or recorded or actual delivery so that there is no doubt that the contractor has received it. This gives the contractor advance warning that the employer may exercise the right to terminate. In many cases, that will be sufficient to stop the default immediately. 'Actual' delivery is delivery by hand. The architect must take care to obtain a receipt acknowledging delivery if this is the chosen method.

Before the architect advises the employer to terminate the contractor's employment by notice the grounds should be considered carefully, as follows.

Wholly or substantially suspends the carrying out of the Works before completion

This ground was broadened in 1994 to include 'substantially suspending'. This overcame the common problem that the contractor has not actually stopped working completely, but for all the good it is doing, it might as well have totally stopped. If the workforce dropped overnight from 40 operatives to just a couple of labourers, termination on this ground may be indicated. It is impossible to give precise guidance and each situation will depend on its own particular facts.

It should be noted that the contractor's action must be without reasonable cause – and a reasonable cause might well be, for example, the employer's failure to make progress payments under the architect's clause 4.3 certificates.

Fails to proceed regularly and diligently with the Works

It has been remarked earlier that it is strange to find this ground for termination included when there is no express requirement for the contractor to proceed regularly and diligently. Nevertheless, the contractor's failure would entitle the employer to terminate under this ground. Whether or not the contractor is proceeding regularly and diligently is a factual question, but one which may be difficult to answer in certain circumstances.

The case of *West Faulkner* v. *London Borough of Newham* (1993) gives excellent guidance on the topic (see section 5.1.2).

Fails to comply with the CDM regulations

It is thought the failure must be clear and unambiguous to justify a termination.

The procedure for terminating the contractor's employment is that if the contractor does not cease its default within seven days of receipt of the default notice, the employer – not the architect – serves a notice by special or recorded or actual delivery, terminating the contractor's employment. The notice operates from the date it is received by the contractor: *J. M. Hill & Sons Ltd* v. *London Borough of Camden* (1980) and clause 6.2.2.

Clause 6.4.2 gives the employer ten days from the expiry of the architect's notice in which to serve the notice of termination. The contract is silent about the situation if the employer fails to terminate within the ten days. The issue of notices of default and termination are usually construed very strictly by the courts: *Robin Ellis Ltd* v. *Vinexsa International Ltd* (2003). Some contracts allow termination without further notice if the default is repeated, but not MW or MWD. It appears, therefore, that if the employer fails to observe the ten-day deadline, the whole procedure, including the service of a default notice by the architect, must be repeated.

Insolvency of contractor

This is the fourth ground. Termination on the ground that the contractor is insolvent is not automatic. It is by notice. The specified grounds are if the contractor (clause 6.1):

- Enters into an arrangement etc. in satisfaction of debts (not being for purposes of amalgamation or reconstruction)
- Passes a resolution or determination for winding up
- Has a winding up or bankruptcy order made
- Has an administrator or administrative receiver appointed
- (If a partnership) each partner is subject to any event noted above.

These are all factual matters, and in some cases, e.g. where a receiver is appointed, it may be best for the architect to advise the employer not to terminate the contractor's employment, since the receiver is probably bound to carry on with the company's contracts. In any event, specialist advice is indicated.

In the case of such insolvency, the contract provides that the employer need only give notice and termination takes effect on receipt of such notice by the contractor. There is no necessity for any warning notice such as is required in the case of a default.

The architect should draft a suitable letter for the employer's signature (Figure 12.3). Termination of the contractor's employment by the employer is governed by an important proviso. The right must not be exercised 'unreasonably or vexatiously' (clause 6.2.1). In simple terms,

Figure 12.3
Letter from employer to contractor terminating employment

SPECIAL OR RECORDED OR ACTUAL DELIVERY

Dear Sir(s)

PROJECT TITLE

Despite the architect's letter to you dated [*insert date of architect's default letter*] you have failed to end your default[1].

In accordance with clause 6.4.2[2] of the conditions of contract, take this as notice that I hereby terminate your employment under this contract without prejudice to any other rights or remedies which I may possess.

You must give up possession of the site of the Works immediately, and should you fail to do so I shall instruct my solicitors to issue appropriate proceedings against you.

Yours faithfully

Copy: Architect

[1] Delete if termination due to insolvency.
[2] Substitute '6.5.1' if determination due to insolvency.

this means that the procedure must not be instituted without sufficient grounds so as merely to cause annoyance or embarrassment. In *J. M. Hill & Sons Ltd* v. *London Borough of Camden* (1980), the Court of Appeal took the view that 'unreasonably' in this context meant 'taking advantage of the other side in circumstances in which, from a business point of view, it would be totally unfair and almost smacking of sharp practice'. In *John Jarvis Ltd* v. *Rockdale Housing Association Ltd* (1986), the word 'unreasonably' in this context was said by the Court of Appeal to be a general term which can include anything which can be judged objectively to be unreasonable, while 'vexatious' connotes an ulterior motive to oppress or annoy.

Corruption

The contract makes express provision if the contractor is guilty of corruption. Essentially, this amounts to the contractor committing an offence under the Prevention of Corruption Acts 1889 to 1916 or, if the employer is a local authority, an offence under section 117(2) of the Local Government Act 1972. In MW 98, the employer's remedy was to 'cancel' the contract. The more precise phraseology under MW and MWD is to be welcomed. All that is necessary is that the employer gives notice of termination. No prior default notice is required. There is no timescale indicated although the wise employer will act promptly.

12.2.2 Consequences of employer termination

Clause 6.7 sets out what is to happen if the employer successfully terminates the contractor's employment under clauses 6.4, 6.5 or 6.6:

- The employer is entitled to engage and pay other contractors to complete the Works. Importantly, clause 6.7.1 expressly provides for both employer and any other contractor to take possession of the site and of the Works. The employer is expressly given permission to use the contractor's temporary buildings, plant, tools, equipment and site materials and, therefore, the contractor may not remove them unless and until instructed to do so. The contractor's licence to occupy the site is at an end and, if the contractor fails to give up possession, it becomes a trespasser in law.
- The employer is relieved of the obligation to make any further payment to the contractor. Clause 6.7.2 expressly states that provisions of the contract which require further payment or release of retention cease to apply.

- Within a reasonable time of completion of the Works and making good of defects the architect must certify or the employer must issue a statement setting out a financial account. In this context, 'completion' can only refer to practical completion under the contract: *Emson Eastern (In Receivership)* v. *EME Developments* (1991). What must be contained in the certificate or statement is clearly listed in clause 6.7.3 as all the expenses incurred by the employer in completing the Works and dealing with defects together with any direct loss and/or damage for which the contractor is liable; the total payments already made to the contractor; and the total amount which would have been payable to the contractor under the terms of the contract.
- Depending on whether the account results in a payment to or from the contractor, clause 6.7.4 stipulates that it is to be a debt payable to or from the contractor.

Clause 6.5.2 sets out some specific consequences of termination resulting from insolvency which, it should be noted, take effect from the date of the insolvency, whether or not the employer has given a termination notice:

- An account is to be prepared in accordance with clauses 6.7.3 and 6.7.4 and the contract clauses requiring further payment or release of retention cease to apply.
- The contractor's obligation to carry out and complete the Works under clause 2.1 is suspended.
- The employer may take reasonable steps to make sure that the site and everything on it are protected and that any materials on site are retained. This last is important, because suppliers may try to remove materials from site for which the contractor has not paid. Whether further payment can be demanded from the employer in this situation is frequently a difficult question of law which depends on all the circumstances.

12.3 *Termination by the contractor*

12.3.1 General

The contractor has a right to terminate its own employment for specified defaults by the employer and if it is successful in doing so, the results for the employer will be disastrous. Because of this, the architect should do everything possible to prevent it from happening.

The practical consequences of a successful contractor termination are:

- The project will have to be completed by others, probably at a higher cost. In any event, professional and other fees will be involved, as well as other expenses.
- The completion date inevitably will be delayed.
- The employer may be faced with a liability to pay the contractor the profit it would have made if the contract had proceeded normally.

Figure 12.4 shows the procedure for termination by the contractor. The grounds and procedure for termination and its consequences are laid down in clauses 6.8, 6.9 and 6.11.

12.3.2 Grounds and procedure

There are six separate grounds which entitle the contractor to exercise its right to terminate its employment under the contract. They are that the employer:

(1) Fails to discharge in accordance with the contract the amount properly due (clause 6.8.1.1)
(2) Interferes with or obstructs any certificate (clause 6.8.1.2)
(3) Fails to comply according to the contract with the CDM Regulations (clause 6.8.1.3)
(4) Causes the whole or substantially the whole of the Works to be suspended for a continuous period of one month before practical completion due to architect's instruction regarding correction of inconsistencies or variations (clause 6.8.2.1)
(5) Causes the whole or substantially the whole of the Works to be suspended for a continuous period of one month before practical completion due to impediment, prevention or default of the employer, architect or any person for whom the employer is responsible (clause 6.8.2.2)
(6) Becomes insolvent in one of the ways specified in clause 6.1 (clause 6.9).

It should be noted that there is no provision that these defaults must be 'without reasonable cause', and the only safeguard from the employer's point of view is that the contractor must not exercise its option to terminate its employment 'unreasonably or vexatiously'.

Figure 12.4
Flowchart of termination by contractor

START

Employer fails to pay amount due by final date of payment?

No

Employer interferes with certificate?

No

Employer fails to comply with CDM Regs?

No

Works suspended for one month due to architect's instruction under 2.4 (2.5 under MWD) or 3.6 or employer's impediment, prevention or default?

Yes

Yes

Yes

Yes

Contractor may issue notice

No

Employer insolvent?

Yes

Yes

Employer stops default or suspension in seven days

No

Yes

Force majeure?

No

Yes

Architect's instructions owing to statutory undertaker's negligence

The consequences of termination are set out in clause 6.11

Contractor may terminate its employment

No

Yes

Specified perils loss or damage?

No

No

Yes

Suspension ends in seven days?

Yes

Civil commotion or terrorism?

No

Employer may give seven-day warning of termination

Yes

Suspension lasts for one month?

Yes

UK Government exercising power

No

No

STOP

As in the case of employer termination, the contractor must follow the procedure precisely as a wrongful termination may amount to a repudiation of the contractor's obligations under the contract. When the contractor wishes to terminate on one of the first five of the above grounds, it must first send a notice to the employer (not to the architect) by special or recorded or actual delivery giving notice of its intention to terminate. That notice must specify the alleged default and require that it ends (Figure 12.5). The employer has seven days from receipt of that notice in which to make good the default, and only if the default is continued for seven days may the contractor terminate its employment. Termination is brought about by a further notice served on the employer by special or recorded or actual delivery and it takes effect on receipt (Figure 12.6).

In the case of the employer's financial failure (clause 6.9.1) no preliminary notice is required.

It is worthwhile considering the grounds for termination in detail because they have been put in to protect the contractor against common wrongdoings by employers.

The employer fails to discharge in accordance with the contract the amount properly due

This ground protects the contractor's right to be paid on time and expressly includes payment of VAT. Steady cash flow is as important to the contractor as it is to the architect or to the employer and in fact this provision is quite generous to the employer. Under clause 4.3 the employer has to pay to the contractor the amounts which the architect certifies as progress payments within 14 days of the date of the architect's certificate and payment is due to the contractor before that period expires.

Any employer who receives a notice under clause 6.8 must pay at once, and if the architect knows about it, the architect should telephone the employer and advise immediate payment and confirm this by a letter along the lines of Figure 12.7.

Right from the outset, the architect must make sure that the employer understands the scheme of payments and the need to pay promptly on certificates. Where possible, financial certificates should be delivered by hand and a receipt obtained. If this is impracticable, they should be sent by special delivery. Once the 14-day period for payment of a certificate has expired, of course, the contractor may refer to adjudication in any case, although many contractors prefer to rely on their option to terminate employment.

Figure 12.5
Letter from contractor to employer giving notice of default or suspension event before termination

SPECIAL OR RECORDED OR ACTUAL DELIVERY

Dear Sir

PROJECT TITLE

We hereby give you notice under clause 6.8.1.1/.2/.3/.4/.5/ [*delete as appropriate*] of the conditions of contract that you are in default in the following respect:

[*insert details of the default or suspension event with dates if appropriate*]

If you continue the default/suspension event [*delete as appropriate*] for seven days after receipt of this notice, we may forthwith terminate our employment under this contract without further notice.

Yours faithfully

Figure 12.6
Letter from contractor to employer terminating employment after default or suspension event notice

SPECIAL OR RECORDED OR ACTUAL DELIVERY

Dear Sir

PROJECT TITLE

We refer to the default/suspension event [*delete as appropriate*] notice sent to you on the [*insert date*].

Take this as notice that, in accordance with clause 7.3.1, we hereby terminate our employment under this contract without prejudice to any other rights or remedies which we may possess.

We are making arrangements to remove all our temporary buildings, plant, etc. and materials from the Works and we will write to you again within the next week regarding financial matters.

Yours faithfully

Figure 12.7
Letter from architect to employer advising immediate payment

Dear Sir

PROJECT TITLE

I refer to our telephone conversation today when I advised you to make immediate payment to the contractor of Certificate No [*insert number*] amounting to £[*insert amount*] in light of the service upon you of a notice preliminary to the contractor terminating its employment under the contract.

You cannot assume that you have a full seven days to pay because, technically, the certificate is not honoured until your cheque has been cleared through the bank. I suggest that you arrange to have your cheque delivered by hand to the contractor's office. If you allow the contractor to terminate its employment the consequences will be considerable extra cost and delay to the project.

It is essential that you pay the certified progress payments within 14 days of the date on the certificate and it is up to you to make the necessary financial arrangements to ensure that funds are available at that time.

Yours faithfully

The cases of *C. M. Pillings & Co Ltd* v. *Kent Investments Ltd* (1985) and *R. M. Douglas Construction Ltd* v. *Bass Leisure Ltd* (1991) establish that if the employer is genuinely disputing the amount of a certificate – and can produce hard evidence to back up the contention – the contractor's action to recover the amount certified may be unsuccessful in arbitration provided the appropriate contractual notices have been served. But it is never sufficient for the employer to refuse to pay until satisfied that it is correct to do so. The architect bears a heavy responsibility to ensure that the certificate is correct.

It is also important to read the wording of the termination provision carefully. It does not say that the contractor may terminate if the employer does not pay the amount certified, but rather refers to the 'amount properly due' in respect of any certificate. This wording would seem to leave open the employer's ordinary equitable and common law rights of set-off providing that the appropriate withholding notices have been served at the right time.

The employer or any person for whom the employer is responsible interferes with or obstructs any certificate

This is a breach of contract at common law and interference or obstruction is a serious matter. In law, conduct of this kind is often referred to as 'acts of hindrance and prevention' and is a breach of an implied term of the building contract. This ground refers to any certificate, not merely financial certificates. There are other certificates that the architect is required to issue which the employer conceivably may try to prevent, such as the practical completion certificate. The architect has a duty under the contract to issue certificates. It must be made plain that the employer who tries to interfere with that duty is in breach. If, despite the warning, the employer absolutely forbids the architect to issue a certificate, the architect is in a difficult position. The architect's duty is then to write and confirm the instructions received, setting out the consequences to the employer (Figure 12.8). The architect has no duty to deliberately inform the contractor, but if the contractor suspects and terminates anyway, the architect has little option but to reveal the facts in any proceedings which may follow. Depending upon the circumstances, it may also be grounds for the architect to terminate his or her engagement by accepting the conduct as repudiation at common law.

Figure 12.8
Letter from architect to employer if employer obstructs issue of a certificate

Dear Sir

PROJECT TITLE

I confirm that a certificate under clause [*insert clause number*] of the conditions of contract is/was [*delete as appropriate*] due on [*insert date*].

I further confirm that you have instructed that I am not to issue this certificate. I am obliged to take your instructions, but you place me in some difficulty and I need to consider my position. The contractor is certain to enquire about the certificate and, if it suspects that you have obstructed the issue, it may exercise its rights to terminate its employment under the contract. There will be serious financial consequences for you.

In the light of the above, I look forward to hearing that you have reconsidered your position.

Yours faithfully

Causes the whole or substantially the whole of the works to be suspended for a continuous period of one month before practical completion due to
- *architect's instructions regarding correction of inconsistencies or variations*
- *impediment, prevention or default of the employer, architect or any person for whom the employer is responsible*

If the carrying out of virtually the whole of the Works is suspended for one month for either of the reasons set out, the contractor may terminate. The first reason relates to the architect issuing instructions regarding the correction of inconsistencies or requiring a variation. If the contractor is delayed for one month, it will be in serious trouble. For any period up to a month, it would be entitled to such loss and/or expense as falls under clause 3.6.3. This clause quite reasonably gives the contractor the option of termination if it foresees no quick end to the suspension and it feels unable to afford to keep the site open.

The second reason is any impediment or default of the employer, the architect, the quantity surveyor or of the employer's persons. The clause unnecessarily emphasises that the delay must not be due to the contractor's own negligence or default. This clause was not present in MW 98.

Fails to comply with CDM regulations

It is thought that the failure must be clear and unambiguous to warrant termination.

Employer's financial failure

These grounds (clause 6.9) parallel those of contractor insolvency in clause 6.5 (see section 12.2.1) and no preliminary notice is required from the contractor. The termination notice must be served by special or recorded or actual delivery. It takes effect on receipt by the employer. It is highly unlikely that the contractor would wish to continue and take its chance of being paid. Clause 6.9.2 provides that, after the occurrence of any of the insolvency events listed in clause 6.1 and even before the notice of termination takes effect, the contractor's obligation to proceed and complete the Works is suspended. This is to avoid the silly situation which would otherwise exist during this period when the contractor would be legally obliged to continue until it could terminate its employment.

Theoretically, this ground is applicable even in the case of a local authority employer, although the whole clause is drafted on the assumption that the employer is an individual or a limited company since, technically, none of the events referred to is applicable in law to local authorities. This is not to say that local authorities cannot get into financial difficulties or become 'insolvent'.

12.3.3 Consequences of contractor termination

If the contractor terminates its employment correctly, it is placed squarely in the driving seat. The contractor, as soon as is reasonably practicable, must prepare an account which must set out:

- The value of work properly executed and materials properly on site for the Works and those for which the contractor is legally bound to pay. The value must be ascertained as if the contractor's employment had not been terminated, together with any other amounts due but not included.
- Any direct loss and/or damage caused by the termination (this will include loss of profit and any other costs caused to the contractor).

The architect is not required to issue a certificate. The contract merely requires the employer to pay to the contractor the amount properly due after taking into account everything previously paid. The payment must be made within 28 days of submission of the contractor's account.

The clause stops short of requiring the employer to pay the balance of the amount on the contractor's account; hence the use of the phrase 'amount properly due'. It is envisaged that the employer will require professional advisers to verify the contractor's account first. However, payment must be made within 28 days and it is no excuse for the employer to plead that the checking process has not been completed. It is suggested the 28 days will not begin to run until the account has been received in a form which, viewed objectively, will allow verification to take place.

Effect of termination on other rights

The contractor's right to terminate its employment is expressly stated to be without prejudice to any other rights or remedies which it may possess. This means that its ordinary rights at common law are preserved, as are those of the employer. This makes it plain that, for example, the contractor can choose if it wishes to terminate the contract under the general law.

12.4 Termination by either employer or contractor

This is new in the Minor Works contract although this type of clause has been included in other contracts such as JCT 98, IFC 98 and ACA 3. Under clause 6.10, either the employer or the contractor may terminate the contractor's employment if the carrying out of the whole or substantially the whole of the uncompleted Works is suspended for a continuous period of one month due to:

- *Force majeure*
- Architect's instructions regarding inconsistencies or variations issued as a result of negligence or default of a statutory undertaking
- Loss or damage to the Works caused by specified perils
- Civil commotion or terrorist activity
- Exercise by the UK Government of statutory power directly affecting the Works.

For either party to operate this clause, the Works must be lacking significant progress for the whole period as a result of the same cause. It is probable, however, that the period must be viewed as a whole.

A seven-day period of notice is required at the end of the suspension of work. The notice must state that unless the suspension is terminated within seven days of receipt of the notice, the employment of the contractor may be terminated. If the suspension does not end, the employer or the contractor may, by further written notice, terminate the contractor's employment.

All the causes of suspension are events beyond the control of the parties. It is likely that both parties will be relieved to bring the contractor's employment to an end in such circumstances.

If the contractor wishes to terminate, however, there is a proviso (clause 6.10.2) to the effect that it is not entitled to give notice if the loss or damage due to specified perils is caused by the contractor's own negligence or default or that of its employees, agents or subcontractors. This proviso is only expressly stating what must be implied – that the contractor must not be able to profit by its own default.

The consequences of termination under clause 6.10 are the same as if the contractor had terminated under clause 6.8 except that the contractor's entitlement to direct loss and/or damage is limited to termination under clause 6.10.1.3 if the loss or damage was due to the negligence or default of the employer or a person for whom the employer is responsible.

12.5 *Summary*

Termination by employer

The employer may terminate the contractor's employment if:

- Without reasonable cause the contractor substantially or wholly suspends work
- The contractor fails to proceed regularly and diligently
- The contractor fails to comply with the CDM Regulations
- The contractor becomes insolvent.

The employer may not exercise this right unreasonably or vexatiously. The procedure laid down must be followed exactly. The architect must advise the employer.

Termination by contractor

The contractor may terminate its own employment if:

- The employer does not pay on time
- The employer interferes with the issue of a certificate
- The employer fails to comply with the CDM Regulations
- The employer suspends the Works for one month
- The employer becomes insolvent.

The contractor must follow the procedure precisely. The common law rights of both parties are preserved.

Termination by either employer or contractor

Either party may terminate if the Works are suspended for one month due to:

- *Force majeure*
- Certain architect's instructions as a result of negligence or default of statutory undertaking
- Loss or damage due to specified perils
- Civil commotion or terrorist activity
- Exercise by UK Government of statutory power.

CHAPTER THIRTEEN
CONTRACTOR'S DESIGNED PORTION (CDP)

13.1 *General*

The contractor had no design responsibility under MW 98 and MW is to the same effect. It has been held that the insertion in the specification of liability for design on the part of the contractor does not override or modify what is in the printed form: *Haulfryn Estate Co Ltd* v. *Leonard J. Multon & Partners and Frontwide Ltd* (1990). However, another court has stated that it would require the clearest possible contractual condition before finding the contractor liable for a design fault: *John Mowlem & Co Ltd* v. *British Insulated Callenders Pension Trust Ltd* (1977). It is certain that giving the contractor design responsibility under MW 98 or MW is not easy.

Overall responsibility for design rests with the architect and the architect can only avoid this responsibility by obtaining the employer's express consent to assigning the responsibility to another: *Moresk Cleaners Ltd* v. *Hicks* (1966). The Minor Works Building Contract with contractor's design 2005 (MWD) does incorporate provisions, albeit they are brief, to enable the contractor to be given design responsibility for specific items. Essentially, the CDP provisions are a very much shortened design and build contract and share some of the features of the DB contract.

13.2 *Documents*

Details of the CDP are to be inserted in the second recital. A footnote advises that a separate sheet may be used if the space is not sufficient to include all the items. The third recital indicates that the employer has had the Employer's Requirements prepared. When the contract is executed and this document is signed by the parties, it becomes part of the contract documents.

There is no provision, as in SBC or ICD, for the contractor to submit formal Contractor's Proposals or for a CDP price analysis.

13.3 *The contractor's obligations*

The contractor's obligations are concisely set out and enlarged in clause 2.1 into which most of the key requirements are packed. Referring to the CDP, the contractor:

- Must complete the design for the CDP including the selection of specifications for materials, goods and workmanship to the extent that they are not stated in the Employer's Requirements
- Must use reasonable skill, care and diligence
- Is not responsible for what is in the Employer's Requirements or for checking the adequacy of any design, but any inadequacy found in the Employer's Requirements must be corrected
- Must comply with the architect's instruction about the integration of the CDP work with the rest of the Works, subject to the contractor's consent under clause 3.4.2
- Must comply with the relevant parts of the CDM Regulations, particularly as they affect the designer
- Must provide the architect with two copies of drawings, details and specification reasonably necessary to explain the CDP as and when necessary to do so.

These provisions are discussed below.

13.4 *Liability*

For what it designed or completed, the contractor would normally have a fitness for purpose liability (*Viking Grain Storage Ltd* v. *T. H. White Installations Ltd* (1985)), but that is modified by clause 2.1.1 to reasonable skill, care and diligence. Therefore, like an architect, the contractor does not guarantee the result of the design, but only that reasonable skill and care was taken in its production. It should be noted that, unlike SBC and ICD, MWD does not expressly refer to the same standard as an architect, but here, unlike the other contracts, that standard is expressly set out.

Clause 2.1.2 deals with existing designs. It is expressly stated that the contractor is not responsible for what is in the Employer's Requirements, nor for verifying whether any design is adequate. This clause is in response to *Co-operative Insurance Society Ltd* v. *Henry Boot Scotland Ltd* (2002) where the court considered an earlier edition of the Contractor's Designed Portion Supplement and held that the contractor had a duty to check any design it was given and to make sure that it worked. The contractor has no duty to check a design that is included in the Employer's Requirements, but if it notices an inadequacy, the Employer's Requirements must be corrected. This clearly includes any defective design. The provision is said to be subject to clause 2.6 which, of course, deals with divergences from statutory requirements. Although the clause does not expressly state that the contractor must notify the architect on finding the inadequacy, it must be implied and, in any event, it is a matter of plain common sense. The correction of the Employer's Requirements must be undertaken by the employer or by the architect on the employer's behalf. Terms would probably be implied that such notifications and corrections will be carried out within a reasonable time. Any delays in correction would entitle the contractor to an extension of time if they delayed the completion date. Problems may occur if an inadequacy exists in the Employer's Requirements, but the contractor completes the design without noticing or checking. Although the contractor specifically has no liability to check, the architect may try to contend that it must have been obvious to the contractor when completing the design. Where the fault lies in these circumstances will be a matter of fact.

Clause 2.2.1 provides that if materials, goods or standards of workmanship in the CDP are not described in the contract documents, they must be of a standard which is appropriate to the Works. Effectively, that means that if there is no description in the Employer's Requirements, the contractor has the responsibility of producing something appropriate.

13.5 Integration of the CDP

Integration with the rest of the Works is a common problem area. The architect is given power to issue directions for integration of the CDP design by clause 2.1.3. The contractor will often contend that the architect's instructions for integration unavoidably result in additional work and will, therefore, seek a variation. The principles are straightforward although application to particular circumstances may need care. There are four possible situations:

- It is a matter for the contractor to allow in the its proposals for the proper integration of the CDP with the rest of the design if the invitation to tender is supported by clear documents.

- The contractor probably has a claim for any additional costs resulting from the architect's directions on integration if the invitation to tender is not supported by sufficient information to enable the contractor to properly design the interface between the CDP and other work.
- The architect must issue directions on integration and the contractor has a claim for additional costs if the architect subsequently issues instructions that affect the CDP.
- The contractor must bear its own costs if it is obliged to alter the CDP in order to correct its own error. In such an instance the architect will probably have to issue some directions about the integration of the corrected CDP.

13.6 *Contractor's information*

Under clause 2.1.5, the contractor is obliged to provide the architect with two copies of design documents reasonably necessary to explain the CDP. The architect is probably entitled to request any related calculations or other information. There are no formal procedures for submission of design information, but there is nothing to stop the architect writing something into the specification if desired. Although not expressly stated, it is likely that the contractor may not proceed with any details until they have been submitted.

Inconsistencies are dealt with in clause 2.5. Clause 2.5.1 adds the Employer's Requirements onto the list of other contract documents. Clause 2.5.2 deals with inconsistencies within the individual CDP documents prepared by the contractor. As might be expected, the contractor must correct the inconsistency at its own cost. However, that is made subject to the architect's satisfaction with the contractor's proposed correction which, it is suggested, must be reasonable and be expressed in writing.

13.7 *Variations*

It is made clear in clause 3.6.1 that an instruction requiring a variation to the CDP can only be issued in respect of the Employer's Requirements. The employer cannot issue an instruction directly about the CDP design. Therefore, it is for the architect to instruct a change to the Employer's Requirements to which the contractor responds by altering its design.

Clause 3.4.2 makes clear that the architect may not issue an instruction which affects the design of the CDP unless the contractor consents. The

consent cannot be unreasonably delayed or withheld. F
wording, it appears that the clause envisages that, if an
struction may affect the design of the CDP, the architect m
in draft to the contractor first. That is because the wording st
architect must not *issue* such an instruction without consent. Therefore, it is not a case of the architect issuing an instruction and the contractor objecting. The contractor is placed firmly in the driving seat.

Valuation of variations in CDP work is to be carried out in accordance with clause 3.6.3 as in MW.

13.8 Other matters

MWD does not require the contractor to take out professional indemnity insurance. Usually, the design will be done by others and they will have ongoing professional indemnity insurance. The employer may need the advice of an insurance broker whether a clause should be written into the contract. If insurance is required, the contractor must maintain it for the appropriate limitation period under the contract: six years if the contract is executed under hand, twelve years if it is a deed.

There is no provision for the employer to receive a licence to use the contractor's design, but it is clear that such a term would be implied.

There are no provisions to limit the contractor's liability nor is there any requirement for as-built drawings in respect of the CDP. It may be thought prudent to require such drawings for future reference.

13.9 Summary

- The contractor has no design liability under the MW contract
- CDP is like a mini design and build contract set in MW
- The contractor must complete the design for the CDP work, but is not responsible for checking any design in the Employer's Requirements
- The architect must issue directions for integration of the CDP with the rest of the Works under certain circumstances
- The contractor's standard is reasonable skill, care and diligence
- The contractor must provide drawings and specification to the architect
- The contractor's consent is required before the issue of an instruction which affects the CDP work
- There is no provision for professional indemnity insurance, limitation of the contractor's liability or as-built drawings.

CHAPTER FOURTEEN
DISPUTE RESOLUTION PROCEDURES

14.1 General

Dispute resolution procedures have been standardised across JCT contracts. In MW and MWD there are four systems.

Little need be said about the first system which is mediation. It is dealt with under clause 7.1. The provisions are brief and simply state that, by agreement, the parties may choose to resolve any dispute or difference arising under the contract through the medium of mediation. The inclusion of this clause seems to be a waste of space. Its redundant nature is shown by the phrase 'The Parties may by agreement . . . '. The parties may do virtually anything by agreement. They could agree to settle disputes by the toss of a coin or the speed of raindrops running down a window pane. In general, there is little point in including as terms of a contract anything which is not already agreed. The whole point of a written contract is that it is evidence of what the parties have already agreed. To have a clause which effectively states 'we may agree to do something else' is pointless. If it is included as a reminder it is best left as a footnote, as it was in MW 98.

In 1996 the Housing Grants, Construction and Regeneration Act (commonly called the Construction Act) (the Act) was enacted (in Northern Ireland Part II of the Act is virtually identical to the Construction Contracts (Northern Ireland) Order 1997). Section 108 of the Act expressly introduces a contractual system of adjudication to construction contracts. Excluded from the operation of the Act are contracts relating to work on dwellings occupied or intended to be occupied by one of the parties to the contract. MW and MWD, in common with other standard forms, incorporate the requirements of the Act. Therefore, all construction Works carried out under these forms are subject to adjudication even if they comprise work to a dwelling house. If MW or MWD is to be used for work to owner-occupied residential property, the adjudication may be deleted if desired. There was some discussion as to whether the adjudication clause would be considered 'unfair' under the provisions of the Unfair Terms in Consumer Contracts Regulations 1999. But it now appears to be established that where the employer enters into a building contract after having been advised by professionals such as architects or quantity surveyors, the adjudication clause will not be considered unfair: *Lovell Projects Ltd* v. *Legg and Carver* (2003); *Westminster Building Company Limited* v. *Beckingham* (2004); *Cartwright* v. *Fay* (2005); and the Court of Appeal decision in *Bryen & Langley* v. *Boston* (2004). Section 108 of the Act provides that:

- A party to a construction contract has the right to refer a dispute under the contract to adjudication
- Under the contract:
  - A party can give notice of intention to refer to adjudication at any time
  - An adjudicator should be appointed and the dispute referred within seven days of the notice of intention
  - The adjudicator must make a decision in 28 days or whatever period the parties agree
  - The period for decision can be extended by 14 days if the referring party agrees
  - The adjudicator must act impartially
  - The adjudicator may use his or her initiative in finding facts or law
  - The adjudicator's decision is binding until the dispute is settled by legal proceedings, arbitration or agreement
  - The adjudicator is not liable for anything done or omitted in carrying out the functions unless in bad faith.
- If the contract does not comply with the Act, the Scheme for Construction Contracts (England and Wales) Regulations 1998 (the Scheme) will apply.

The right to refer to adjudication 'at any time' means that adjudication can be commenced even if legal proceedings (and presumably arbitration) are in progress about the same dispute: *Herschel Engineering Ltd* v. *Breen Properties Ltd* (2000). It has also been held that adjudication can be sought even if repudiation of the contract has taken place. A dispute may be referred to adjudication and the adjudicator may give a decision even after the expiry of the contractual limitation period: *Connex South Eastern Ltd* v. *M. J. Building Services plc* (2005). In such a case, the referring party runs the risk that the respondent will use the limitation period defence, in which case the claim will usually fail.

Adjudication is dealt with in Article 6 and clause 7.2. It is sometimes referred to as a temporarily binding solution. Experience suggests that, in most instances, parties tend to accept the adjudicator's decision as final. Where there are challenges through the courts against the enforcement of an adjudicator's decision, the challenge can only be concerned with matters such as the adjudicator's jurisdiction or whether the adjudicator complied with the requirements of natural justice, not whether the adjudicator's decision itself was correct. The courts cannot interfere with the adjudicator's decision, no matter how obviously wrong: *Bouygues UK Ltd* v. *Dahl-Jensen UK Ltd* (2000). It is sometimes said that it does not matter if the adjudicator answers the right question in the wrong way, but an adjudicator who answers the wrong question, even in the right way, will lack jurisdiction.

The parties must comply with the adjudicator's decision, following which a dissatisfied party may take the dispute to arbitration or legal proceedings depending upon which system has been chosen in the contract. The parties are not appealing against the decision of the adjudicator, they are starting from the beginning as though the adjudication had not taken place. The arbitrator or court will ignore the adjudicator's previous decision and reasoning in arriving at an award or judgment respectively: *City Inn Ltd* v. *Shepherd Construction Ltd* (2002).

If the parties wish to have a final and binding decision rather than submit a dispute to adjudication, they have a choice between arbitration and legal proceedings. It should be noted that, unlike the position under previous contracts up to and including MW 98, legal proceedings will apply unless the contract particulars are completed to show that arbitration is to be the procedure. The advantages of arbitration are said to be:

- *Speed*: A good arbitrator should dispose of most cases in months, not years.
- *Privacy*: Only the parties and the arbitrator know the details of the dispute and the award.
- *The parties decide*: The parties can decide time-scales, procedure and whether to have a hearing.
- *Expense*: The speed and technical expertise of the arbitrator ought to keep costs down. The arbitrator can cap the costs.
- *Technical expertise of the arbitrator*: A technical arbitrator may shorten the time and may avoid the need for expert witnesses.
- *Appeal*: The award is usually final.

Disadvantages may be:

- In theory, it is more expensive because the parties pay the cost of the arbitrator and the hire of a room.
- A poor arbitrator can slow the process and create expense.
- The arbitrator may not be an expert on the law.
- The parties may not be satisfied with the arbitrator.

The advantages of legal proceedings are said to be:

- The judge is an expert on the law.
- The Civil Procedure Rules make for an efficient and speedy process.
- Several parties can be joined into the proceedings.
- Low costs of judge and courtroom.
- There is always the possibility of an appeal.

The disadvantages of legal proceedings are said to be:

- Few judges know much about construction.
- Parties cannot choose the judge.
- Costs will be added because expert witnesses will be needed to assist the judge.
- Cases often drag on for a long time.
- The parties' own costs may be high as a result of long time-scales.
- A party may be forced into the appeals process with resultant unacceptable levels of legal costs.

Arbitration is probably still the most satisfactory procedure for the resolution of construction disputes and employers would be advised to complete the contract particulars accordingly. Where the parties have agreed that the method of binding dispute resolution will be arbitration, a partly who attempts to use legal proceedings instead will fail in a costly way if the other party calls in aid section 9 of the Arbitration Act 1996. Section 9 requires the court to grant a stay (postponement) of legal proceedings until the arbitration is concluded unless the arbitration is null, void, inoperable or incapable of being performed. The court has no discretion about the matter and the successful party will claim its costs. The result is not only that the party intent on legal proceedings will have to revert to arbitration, but it will have to pay the other party's legal costs in opposing the legal proceedings.

Arbitration is dealt with by Article 7, clause 7.3 and schedule 1. Legal proceedings are dealt with by Article 8.

14.2 *Adjudication*

14.2.1 The contract provisions

Article 6 provides that, if any dispute or difference arises under the contract, either party may refer it to adjudication in accordance with clause 7.2. The parties are of course the employer and the contractor. It is likely to be the contractor which initiates adjudication, but there is nothing to stop the employer from doing so. Among other things, the employer could seek adjudication if the architect appears to be too generous in certification of money or if the contractor refuses to rectify a serious defect in the Works.

It should be noted that only disputes arising 'under' the contract may be referred. An adjudicator has no jurisdiction to deal with disputes about any other agreements made by the parties even if they were made in relation to the contract of which the adjudication clause forms part: *Shepherd Construction* v. *Mecright Ltd* (2000). Very often the parties will enter into an agreement to settle some intermediate matter which then later becomes the subject of a dispute. Occasionally, parties will enter into a so-called acceleration agreement which is ancillary to the main contract. None of these fall under the adjudicator's jurisdiction, which is very narrowly circumscribed.

Adjudication is optional. It is not necessary before seeking arbitration or legal proceedings. Nevertheless, it is rapidly replacing arbitration as the standard dispute resolution process. Unfortunately, adjudication is often used for complex disputes involving large amounts of money, for which it is not suited.

Clause 7.2 is much shorter than the comparable clause in MW 98. That is because the procedure set out in MW 98 has been abandoned in favour of the procedure in the Scheme. This is sensible, because the Scheme is a comprehensive set of rules especially drafted to comply with the Act. Currently, there are too many procedures put forward by many and varied organisations.

Use of the Scheme is made subject to the proviso that the adjudicator and nominating body are to be those stated in the contract particulars.

The architect cannot be the respondent to an adjudication under MW or MWD, because the architect is not a party to the contract. Architects can act as witnesses and give evidence, whether in person at a hearing or by means of a witness statement, but they have no duty to run an adjudication on behalf of the employer. Indeed, it would be foolish for an architect to do so unless very experienced in such matters. Acting in an adjudication usually calls for a degree of skill and experience which most construction professionals, acting in the normal course of their professions, will not readily acquire. If the dispute is other than very straightforward or where one party has retained the services of a legal representative, the other party is well advised to do likewise.

The main provisions of the Scheme are explained below.

14.2.2 The Scheme: starting the adjudication process

Paragraph 1 of the Scheme states that any party to a construction contract may give to all the other parties a written notice of an intention to refer a dispute to adjudication. The notice must describe the dispute and who are the parties involved. It must give details of the date and location, the redress sought and the names and addresses of the parties to the contract. An example of such a notice is shown in Figure 14.1.

The notice starts the adjudication process and it is also an important document in its own right. Great care must be taken in its preparation because the dispute which the adjudicator is entitled to consider is the dispute identified in the notice of adjudication: *McAlpine PPS Pipeline Systems Joint Venture* v. *Transco plc* (2004). The dispute cannot be broadened later by the referring party, although it can be elaborated and more detail provided: *Ken Griffin and John Tomlinson* v. *Midas Homes Ltd* (2000). The purpose of the notice is to tell the other party the nature of the dispute, to give the same information to the people responsible for nominating the adjudicator and importantly, of course, to define the dispute, to specify exactly the redress sought, and the parties concerned. The adjudicator can only answer the question posed in the notice of adjudication. The adjudicator is not empowered to answer the question which should have been asked and an adjudicator who tried to do that would be acting in excess of jurisdiction. The decision would be a nullity.

Paragraph 20 of the Scheme allows the adjudicator to take into account any other matters which the parties agree should be within the adjudication's scope. Moreover, the adjudicator is expressly empowered to take into account matters which the adjudicator considers are necessarily connected with the dispute. The express empowerment merely puts into words what would be the legal position in any event: *Karl Construction (Scotland) Ltd* v. *Sweeney Civil Engineering (Scotland) Ltd* (2002); *Sindall Ltd* v. *Solland* (2001).

Figure 14.1
Letter from one party to the other giving notice of referral

SPECIAL/RECORDED DELIVERY

Dear Sir

PROJECT TITLE

I/We hereby give notice that it is my/our intention to refer the following dispute or difference to adjudication in accordance with Article 6 of the contract between us dated [*insert date*].

The dispute is:

[*Specify the dispute sufficiently so that the other party can readily identify it, but without setting out the detailed grounds of the case. For example: 'Failure of the architect to properly value an instruction under the provisions of clause 3.6.' The redress should also be set out*]

The date of the dispute is [*insert date or range*] and the location is [*specify the location*].

The parties to the dispute are [*specify names and addresses*] who are also the parties to the contract [*if they are not the parties to the contract, state who they are and why there is a right to adjudication*].

Yours faithfully

14.2.3 The Scheme: appointment of the adjudicator

Paragraphs 2 to 6 set out the procedure for selecting an adjudicator. Paragraph 2 contains the framework and it is made subject to the overriding point that the parties are entitled to agree the name of an adjudicator after the notice of adjudication has been served. If the parties can so agree, they have the best chance of an adjudicator who has the confidence of both parties. If there is a person named in the contract, that person must first be asked to act as adjudicator. There are problems associated with naming a person in the contract: the person named may be unavailable to act when called upon or the person's expertise may be unsuitable for the particular dispute.

In the event that there is no person named in the contract or if that person is not available for some reason, the referring party must ask the nominating body indicated in the contract particulars to nominate an adjudicator. If the referring party does ask for a nomination, both parties are stuck with the result, even if the adjudicator is quite poor, unless they jointly agree to revoke the appointment and start again.

The nominating body is to be stated in the contract particulars. The bodies listed are:

- Royal Institute of British Architects
- Royal Institution of Chartered Surveyors
- Construction Confederation
- National Specialist Contractors Council
- Chartered Institute of Arbitrators.

If there is no adjudicator named in the contract particulars and no body is selected, the referring party may choose any one of the bodies to make the nomination. If all the nominating bodies have been deleted, the referring party is free to choose any nominating body at all to make the appointment. A nominating body is fairly broadly defined in paragraph 2(3) as a body which holds itself out publicly as a body which will select an adjudicator on request.

Paragraph 6 states that if an adjudicator is named in the contract, but for some reason cannot act or does not respond, the referring party has three options. The first is to ask any other person specified in the contract to act. The second is to ask the adjudicator-nominating body in the contract to nominate and the third is to ask any other nominating body to nominate. It will readily be seen that this procedure is simply a clarification of existing options.

The nominated adjudicator must accept within two days from receiving the request. The adjudicator must be a person and not a body corporate. Therefore, a firm of quantity surveyors could not be nominated although one of the directors or partners of that firm could be nominated. Paragraph 5 gives the nominating body five days from receipt of the request to give the nomination to the referring party and presumably also to the respondent although the Scheme does not expressly say so. If the nominating body does not nominate within five days, the parties are free to agree on the name of an adjudicator or the referring party may request another nominating body to nominate. In either case the chosen adjudicator has two days to respond as before.

It is sometimes assumed that if a party does not wish the adjudication to proceed, it is enough simply to lodge an objection with the nominator in order to stop the process. The nominating body will not be interested. Once it has nominated an adjudicator, its job is finished. Paragraph 10 of the Scheme states that the objection of either party to the adjudicator will not invalidate the appointment nor any decision reached by the adjudicator. Of course, it is a matter for each party whether or not, and the extent to which, it wishes to take part in the adjudication. A party objecting cannot stop the adjudication proceeding and it is usually sensible to take part. After all, the objecting party may even win the adjudication rendering the objection academic. But, in taking part, it should make clear that participation is without prejudice to its objection and to the party's right to refer the objection to the courts in due course. If a party, knowing of grounds for objection, says nothing but continues the adjudication without objection, it may be deemed to have accepted the adjudicator: *R. Durtnell & Sons Ltd* v. *Kaduna Ltd* (2003).

If the adjudicator resigns or the parties revoke the appointment, the position is covered by paragraphs 9 and 11. The adjudicator may resign at any time on giving notice in writing to the parties. There is no requirement that the notice should be reasonable and immediate resignation is possible. The referring party may serve a new notice of adjudication and seek the appointment of a new adjudicator. If the new adjudicator requests it and if reasonably practicable, the parties must make all the documents available which have been previously submitted. Sometimes it is clear that the dispute referred is the same as a dispute which has already been the subject of an adjudication decision. In such circumstances, the adjudicator must resign. The adjudicator is entitled to determine a reasonable amount due in fees and expenses and how it is to be apportioned between the parties. They are jointly and severally liable for any outstanding sum. Although it is not stated as a reason for resignation, the significant variation of a dispute from what was contained in the referral notice, so that the adjudicator is not competent to decide it, is also a ground for payment. This appears to refer to a significant difference between the dispute set out in the notice of adjudication and the dispute included in the referral.

It will probably be a rare occasion if the parties agree to revoke the appointment under paragraph 11. If they do, the adjudicator is entitled to determine a reasonable amount of fees and expenses and the apportionment. The parties, as before, are jointly and severally liable for any balance. It seems that the amount of fees determined by an adjudicator cannot be challenged unless the adjudicator can be shown to have acted in bad faith. It is not sufficient that a court would have arrived at a different figure: *Stubbs Rich Architects* v. *W. H. Tolley & Son* (2001). However, it may be that an adjudicator would not be entitled to a fee if a court decided that the decision was not enforceable by reason of a serious breach of the rules of natural justice. This point was floated, but did not have to be decided, in *Dr Peter Rankilor & Perco Engineering Service Ltd* v. *M. Igoe Ltd* (2006).

14.2.4 The Scheme: the adjudication process

Paragraph 3 of the Scheme states that a request for the appointment of an adjudicator must include a copy of the notice of adjudication. There is not much time available for the nomination because paragraph 7 stipulates that the referring party must submit the dispute in writing to the adjudicator (known as the 'referral notice'), with copies to each party to the dispute, no later than seven days after the notice of adjudication. On receipt of the name of the adjudicator, the referring party must be ready to despatch the referral notice almost immediately. In practice, a referring party will have the referral notice ready at the time the notice of adjudication is issued. The referral notice, which is the referring party's claim, must be accompanied by relevant parts of the contract and whatever other evidence the referring party relies upon in support of the claim.

To comply with the rules of natural justice, and although the Scheme does not expressly state that the respondent may reply to the referral notice, the adjudicator must allow a reasonable period for the reply. It is now a matter for the adjudicator to set the period for the respondent's reply, but in view of the restricted overall period in which the decision must be made, 14 days is the very most which any respondent can expect. In most instances, the adjudicator will allow between seven and ten days.

According to paragraph 19, the adjudicator must reach a decision 28 days after the date of the referral notice. The period may be extended by 14 days if the referring party consents or for a longer period if both parties agree. Paragraph 19(3) stipulates that the adjudicator must deliver a copy of the decision to the parties as soon as possible after the decision has been reached. If the adjudicator does not comply with this timetable, either party may serve a new notice of adjudication and request a new adjudicator to act. The new adjudicator may request copies of all documents given to the former adjudicator.

Occasionally, an adjudicator does not arrive at a decision within the required time-scale or does not immediately send it to the parties. There are two English decisions about this situation: *Simons Construction Ltd* v. *Aardvark Developments Ltd* (2003) and *Barnes & Elliott Ltd* v. *Taylor Woodrow Holdings Ltd* (2004). The position appears to be as set out in those decisions: a late or late-communicated decision is valid provided neither party has taken steps to bring the adjudication to an end, such as serving a fresh notice of adjudication after the expiry of the relevant period.

If one of the parties fails to comply with the adjudicator's decision, the other may seek enforcement of the decision through the courts. The courts will normally enforce the decision unless there is a jurisdictional or procedural problem. In enforcement proceedings, the court is not being asked to comment on the adjudicator's decision or reasoning, although a court will quite often do so, thus obscuring the *ratio* of the judgment. Where a court is asked to enforce an adjudicator's decision, the important part of the judgment is simply the reason why the judge decided to enforce or not. Any comments the judge may make on the adjudicator's decision itself will be *obiter;* at best persuasive, but certainly not of binding force.

14.2.5 The Scheme: important powers and duties of the adjudicator

Although there are still some adjudicators who believe that they are entitled to reach their decisions on the basis of their ideas of fairness, moral rights or justice, it is clear that, under paragraph 12 of the Scheme, the adjudicator's duties are to act impartially in accordance with the relevant contract terms, to reach a decision 'in accordance with the applicable law in relation to the contract' and to avoid unnecessary expense. Elsewhere, the adjudicator's duty has been stated to be 'primarily to decide the facts and apply the law (in the case of an adjudicator, the law of the contract)' (*Glencot Development & Design Company Ltd* v. *Ben Barrett & Son (Contractors) Ltd* (2001)).

The Scheme gives the adjudicator powers which are both very broad and precise:

- To take the initiative in ascertaining the facts and the law
- To decide the procedure in the adjudication
- To request any party to supply documents and statements
- To decide the language of the adjudication and order translations
- To meet and question the parties
- To make site visits, subject to any third party consents
- To carry out any tests, subject to any third party consents
- To obtain any representations or submissions
- To appoint experts or legal advisers, subject to giving prior notice
- To decide the timetable, deadlines and limits to length of documents or oral statements
- To issue directions about the conduct of the adjudication.

Under paragraph 8, the adjudicator may adjudicate at the same time on more than one dispute under the same contract, if all the parties concerned give their consent. With the consent of all, the adjudicator may deal with related disputes on several contracts even if not all the parties are parties to all the disputes. The parties may agree to extend the period for decision on all or some of the disputes. Multiple dispute procedures can become quite complicated. Where there are different contracts and the parties vary from one contract to another, it will be a matter of discussion and agreement whether separate adjudications should be conducted at the same time. Under paragraph 14, the parties must comply with the adjudicator's directions. If one of the parties does not comply, the adjudicator, under paragraph 15, has the power to continue the adjudication notwithstanding the failure, to draw whatever inferences the adjudicator believes are justified in the circumstances or to make a decision on the basis of the information provided and to attach whatever weight to evidence submitted late that the adjudicator thinks fit.

A party may have assistance or representation as deemed appropriate, but oral evidence or representation may not be given by more than one person unless the adjudicator decides otherwise (paragraph 16).

Sometimes a contract stipulates that a decision or certificate is final and conclusive. Except in those circumstances, where the decision is stated to be both final *and* conclusive, under paragraph 20 the adjudicator is given power to open up, revise and review any decision or certificate given by a person named in the contract. Therefore, it appears that a contract which simply states that a certificate is conclusive is open to review. On that basis, the final certificate in SBC is not open to review because it is called 'final' and stated to be conclusive. However, under MW and MWD the final certificate is not stated to be conclusive and it is, therefore, open to review. The adjudicator can order any party to the dispute to make a payment, decide its due date and the final date for payment and decide the rates of interest, the periods for which it must be paid and whether it must be simple or compound interest. In deciding what, if any, interest must be paid, the adjudicator must have regard to any relevant contractual term. To 'have regard' to a contractual term is a rather loose phrase which probably means little more than to give attention to it.

Paragraph 17 makes clear that the adjudicator must consider relevant information submitted by the parties and if the adjudicator believes that other information or case law should be taken into account, it must be provided to the parties and they must have the opportunity to comment: *RSL (South West) Ltd* v. *Stansell Ltd* (2003). Paragraph 18 requires the parties, including the adjudicator, not to disclose to third parties information which the supplier has said should be treated as confidential, unless the disclosure is necessary for the adjudication.

14.2.6 The Scheme: the decision

The adjudicator may order the parties to comply peremptorily with the whole or any part of the decision (paragraph 23). If the adjudicator does not give any directions about compliance, the parties must comply immediately they receive the decision (paragraph 26). The decision will be binding and the parties must comply until the dispute is finally determined by arbitration, legal proceedings or by agreement.

The adjudicator must give reasons for the decision if either party requests them. If reasons are requested, limited reasons or, as some adjudicators say, 'an indication of the factors influencing the decision', are not sufficient and such statement has little if any practical effect. The reasons are to be read with the decision and may be used as a means of construing and understanding the decision: *Joinery Plus Ltd (in administration)* v. *Laing Ltd* (2003). If the adjudicator is asked for comments about the decision after it has been delivered, they are irrelevant except to the extent that the adjudicator is entitled to correct basic mistakes in the decision, if invited to do so: *Bloor Construction (UK) Ltd* v. *Bowmer & Kirkland (London) Ltd* (2000).

The adjudicator may decide on reasonable fees. The parties are jointly and severally liable for payment if the adjudicator makes no apportionment or if there is an outstanding balance. Paragraph 26, following the Act, makes clear that the adjudicator will not be liable for anything done or omitted in carrying out the functions of an adjudicator unless the act or omission is in bad faith. The same protection is given to any employee or agent of the adjudicator. However, this paragraph is only binding on persons who are parties to the contract.

14.2.7 The Scheme: award of costs

There is nothing in the Scheme which allows the adjudicator to award the legal costs of the parties. The Act does not encourage the parties to incur large costs in pursuing claims. However, difficulties have arisen over this point in some instances. It now seems clear that although there is no provision in the Scheme which gives the adjudicator power to award costs, the adjudicator can be given such power, either expressly by the parties or by implied agreement: *Northern Developments (Cumbria) Ltd* v. *J. & J. Nichol* (2000).

14.3 *Arbitration*

14.3.1 General

It used to be thought that a court did not have the same power as the arbitrator to open up and revise certificates and decisions. That approach was changed by *Beaufort Developments (NI) Ltd* v. *Gilbert-Ash NI Ltd* (1998) which held that a court has the same powers as an arbitrator to open up, review and revise certificates, opinions and decisions of the architect. The court has this power as a right, whereas in arbitration, the power must be conferred upon the arbitrator by the parties to the contract.

Doubtless in the past, parties chose arbitration so that they could be sure of being able to refer, for scrutiny, certificates and other decisions of the architect if the need arose. Now that it has been established that courts have this power, it is not essential to choose arbitration as the dispute resolution procedure. However, there are other factors which may still persuade the parties that arbitration is more suited to construction disputes and these factors have been listed at the beginning of this chapter. Note that, unless arbitration is expressly chosen, legal proceedings are the default option.

Arbitration can take place on any matter at any time: *Crown Estates Commissioners* v. *John Mowlem & Co Ltd* (1994). It is almost always costly and time-consuming. The outcome, no matter how strong the case appears to be, will be uncertain. Often the successful party will, with hindsight, consider that the expenditure of cost, time and effort was not worth the result achieved. It must always be remembered that even successful parties will not recover more than 60–80% of their costs.

Statistics seem to indicate that there is a general reduction in arbitrations in favour of adjudication, even though they are not strictly comparable in effect. But there will always be some people who will threaten arbitration over a small matter in an attempt to gain an advantage, perhaps in a negotiation situation where the ploy of abandoning talks and serving an arbitration notice is still practised. Sadly, even the introduction of the adjudication process will not entirely banish that approach. It is not always possible to avoid arbitration and, therefore, employers and contractors must ensure that they properly appreciate how the process operates. Only then can they recognise the possible consequences.

It is commonly thought that the arbitration process is fairly informal – an opportunity for the warring parties to get together to enable them to chat to the arbitrator about the matters in dispute before the arbitrator, with the parties, agree on the way forward. That is not what arbitration is about. The novice will discover that arbitration has evolved into a very formal procedure, very much like court proceedings in some ways although carried on in private. After the arbitration has been commenced by the service of the formal arbitration notice, the first thing is the 'preliminary meeting'. The arbitrator invites the parties, but it is not a friendly discussion. Rather, it is a formal meeting to establish all the important things which need to be decided before the arbitration can progress. It is usual for the arbitrator to have an agenda. It is not unknown for one of the parties to attempt to gain an advantage by springing a surprise request on the arbitrator at that meeting. A party who goes to a preliminary meeting without taking a fully briefed legal adviser experienced in arbitration is simply asking for trouble.

The employer and contractor have the right to agree who should be appointed, or should appoint, the arbitrator and they are free to agree important matters such as the form and timetable of the proceedings. Provided the parties are sensible and prepared to co-operate to some degree, this raises the possibility of a quicker procedure than would otherwise be the case in litigation and even matters such as the venue for any future hearing might be arranged to suit the convenience of the parties and their witnesses.

MW arbitration procedures are set out in schedule 1 of the contract. The JCT 2005 edition of the Construction Industry Model Arbitration Rules (CIMAR) current at the contractual base date are to govern the proceedings, (schedule 1, paragraph 1). The provisions of the Arbitration Act 1996 are expressly stated by paragraph 6 to apply to any arbitration under this agreement. That is to be the case no matter where the arbitration is conducted. Therefore, even if the project and the arbitration take place in a foreign jurisdiction, the UK Act will apply provided that the parties contracted on MW or MWD, and clause 1.7 referring to the law of England is not amended.

Article 7 expressly excludes from the arbitration procedure under the contract: disputes about value added tax, disputes under the Construction Industry Scheme, where legislation provides some other method of resolving the dispute, and the enforcement of any decision of an adjudicator.

By paragraph 5 in accordance with sections 45(2)(a) and 69(2)(a) of the Act, the employer and contractor agree that either party may, by proper notice to the other and to the arbitrator, apply to the courts to determine any question of law arising in the course of the reference and appeal to the courts on any question of law arising out of an award. When the clause was originally introduced, there was doubt whether the courts would accept it as satisfying the requirements of such an appeal. However, such clauses have been held to be effective: *Vascroft (Contractors) Ltd* v. *Seeboard plc* (1996).

14.3.2 The appointment of an arbitrator

Either party may begin arbitration proceedings. As a first step, one party must write to the other requesting them to concur in the appointment of an arbitrator (paragraph 2.1). Whoever does so, proceedings are formally commenced when the written notice is served. Rule 2.1 of CIMAR sets out the procedure, stating that the notice must identify the dispute and require agreement to the appointment of an arbitrator. Figure 14.2 is a suitable letter. The party seeking arbitration normally offers the names of three prospective arbitrators. This saves time and often both parties can agree on one of the names. If the respondent's position is that none of the names are acceptable, it is usual for the dissenting party to volunteer a new set of names. The arbitrator must have no relationship to

Figure 14.2
Letter from one party to the other requesting concurrence in the appointment of an arbitrator

SPECIAL/RECORDED DELIVERY

Dear Sir

PROJECT TITLE

I/We hereby give you notice that we require the undermentioned dispute(s) or difference(s) between us to be referred to arbitration in accordance with Article 7 of the contract between us dated [*insert date*]. Please treat this as a request to concur in the appointment of an arbitrator under paragraph 2.1 of schedule 1.

The dispute(s) or difference(s) is/are:

[*Specify*]

I/We propose the following three persons for your consideration and require your concurrence in the appointment of one of them within 14 days of the date of service of this letter, failing which I/we shall apply to the President of the [Royal Institute of British Architects, Royal Institution of Chartered Surveyors or Chartered Institute of Arbitrators as appropriate] for the appointment of an arbitrator under Article 7.

The names and addresses of the persons we propose are:

(1) Sir Bertram Twitchett, RIBA, FCIArb, Fawlty Towers, Probity, Hants., etc.

Yours faithfully

either of the parties nor should the arbitrator have connections with any matter associated with the dispute.

The importance of agreeing on a suitable candidate rather than having one appointed whose skills and experience may be entirely unknown cannot be overemphasised. Although there may be little that the parties can agree, they should make a sincere effort to agree on the arbitrator. If no agreement can be reached, paragraph 2.1 of the contract and rule 2.3 of CIMAR provide that if the parties cannot agree upon a suitable appointment within 14 days of a notice to concur or any agreed extension to that period, either party can apply to a third party to appoint an arbitrator. There is a list of appointors in the contract particulars against schedule 1. All but one should be deleted, leaving the agreed appointor as either: the President or a Vice-President of the Royal Institute of British Architects, the Royal Institution of Chartered Surveyors, or the Chartered Institute of Arbitrators. If no single body has been chosen the default provision is the President or a Vice-President of the Royal Institute of British Architects. Figure 14.3 is the sort of letter that a contracting party might send to an appointing body requesting appointment.

It is possible for the parties to the contract to insert the name of a different appointor of their choice at the time the contract was executed. However, it will be necessary to complete special forms and to pay the relevant fee. Although the system of appointing an arbitrator varies, the aim is the same. The object is to appoint a person of integrity who is independent, having no existing relationships with either party or their professional advisers and who is impartial. It should go without saying that the arbitrator should have the necessary and appropriate technical and legal expertise. Claimants who have a dispute to refer and respondents receiving a notice to concur should waste no time in taking proper expert advice on how best to proceed.

If the arbitrator's appointment is made by agreement, it will not take effect until the appointed person has confirmed willingness to act, irrespective of whether terms have been agreed. If the appointment is the result of an application to the appointing body, it becomes effective, whether or not terms have been agreed, when the appointment is made by the relevant body (CIMAR rule 2.5). There is no fixed scale of charges for arbitrator's services and fees ought to depend on their experience, expertise and often on the complexity of the dispute. Arbitrators commonly charge substantial fees. They usually require an initial deposit from the parties and, if there is to be a hearing, there will be a cancellation charge graded in accordance with the proximity of the cancellation to the start of the hearing. The argument in support of this is that the arbitrator will have put one or two weeks on one side for the hearing during

Figure 14.3
Letter from architect on employer's behalf or from contractor to appointing body

Arbitration Appointments Office,
Royal Institute of British Architects,
66 Portland Place,
London W1B 1AD
[or alternative]

Dear Sir

PROJECT TITLE

[I am acting as architect for [*name of employers*] *or* We are the contractor] under a contract in MW [*or MWD*] Form, Article 7 of which makes provision for your President to appoint an arbitrator in default of agreement.

Will you please send me/us the appropriate form of application and supporting documentation, together with a note of the current fees payable on application.

Yours faithfully

which time no other work has been booked. A cancellation means that it is difficult for the arbitrator to secure work at short notice to fill the void. In cases where the cancellation fee is large, due to proximity to the hearing date, it might be sensible to ask the arbitrator to account to the parties for activities during the hearing period.

After appointment, the arbitrator will consider which of the procedures summarised appears to be most appropriate as a forum for the parties to put their cases. The arbitrator must choose the format that will best avoid undue cost and delay and that is often a most difficult balancing act. Therefore, within 14 days of the acceptance of the appointment being notified, the parties must provide the arbitrator with an outline of their disputes and of the sums in issue, along with an indication of which procedure they consider best suited to them. After due consideration of all parties' views and unless a meeting is considered unnecessary, the arbitrator must, within 21 days of the date of acceptance, arrange a meeting (the preliminary meeting), which the parties or their representatives will attend, to agree (if possible) or receive the arbitrator's decision upon everything necessary to enable the arbitration to proceed. It is obviously preferable for the parties to agree which procedure is to apply. If they cannot agree, the documents-only procedure will apply unless the arbitrator, after having considered all representations, decides that the full procedure will apply.

Although the parties may conduct their own cases if they so wish, it is usually better to engage experienced professionals to act for them, because the proceedings can become very formalised and the inexperienced can be severely disadvantaged despite the best efforts of the arbitrator.

14.3.3 Powers of the arbitrator

Arbitrators appointed under the arbitration provisions of MW or MWD are given very wide express powers. Their jurisdiction is to decide any dispute or difference of any kind whatsoever arising under the contract or connected with it (Article 7). The scope could scarcely be broader (*Ashville Investments Ltd* v. *Elmer Contractors Ltd* (1987) and by paragraph 3 of schedule 1, the arbitrator's powers embrace:

- Rectifying of the contract so that it accurately reflects the true agreement between the parties
- Directing the taking of measurements or the undertaking of such valuations as the arbitrator thinks desirable to determine the respective rights of the parties

- Ascertaining and awarding any sum which should have been included in a certificate
- Opening up, reviewing and revising any certificate, opinion, decision, requirement or notice and determining all matters in dispute as if no such certificate, opinion, decision, requirement or notice had been given.

The 1996 Arbitration Act broadened the arbitrator's powers. For example, an arbitrator may:

- Order which documents or classes of documents should be disclosed between, and produced by, the parties: section 34(2)(d)
- Order whether the strict rules of evidence shall apply: section 34(2)(f)
- Decide the extent to which the arbitrator should take the initiative in ascertaining the facts and the law: section 34(2)(g)
- Take legal or technical assistance or advice: section 37
- Order security for costs: section 38
- Give directions in relation to any property owned by or in the possession of any party to the proceedings which is the subject of the proceedings: section 38
- Make more than one award at different times on different aspects of the matters to be determined: section 47(1)
- Award interest: sections 49(1) to 49(6)
- Make an award on costs of the arbitration between the parties: sections 61(1) and 61(2)
- Direct that the recoverable costs of the arbitration, or any part of the arbitral proceedings, are to be limited to a specified amount: sections 65(1) and 65(2).

14.3.4 CIMAR procedure

If an arbitration is commenced under the provisions of MW or MWD, subject to the law of England, it must be conducted in accordance with the JCT 2005 edition of CIMAR which is current at the base date of the contract. If any amendments have been issued by JCT since that date, the parties may, by a joint written notice to the arbitrator, state that they wish the reference to be conducted according to the amended rules (paragraph 1). CIMAR is a clear and comprehensive body of rules. There are two appendices, the first defining terms and the second helpfully reproducing relevant sections of the Arbitration Act 1996 which are relevant, but not already included in the rules. These are followed by the JCT Supplementary and Advisory Procedures, the first part of which is mandatory and must be read with the rules. The greater part is advisory only, but appears to be well worth adopting. At the back of the document is a set of notes prepared by the Society of Construction Arbitrators dated 1 February 1998, updated January 2002. The whole document, at the time of writing, is available on www.jctcontracts.com. JCT/CIMAR is very detailed and repays careful study.

If it is necessary to present oral evidence, cross-examination will be required and this is usually done at a hearing. Hearings are the private equivalent of a trial. They are conducted in private. It has already been noted that parties may represent themselves if they wish, but most are represented by an experienced professional who ideally is both legally and technically qualified. There is no requirement to be represented by solicitor and counsel and indeed it is sometimes argued that litigation solicitors, with their detailed knowledge of court procedure, fit badly into an arbitration scenario where there is the opportunity for the parties to move relatively quickly and agree less formal procedures. JCT/CIMAR offers the parties a choice of three broad categories of procedure by which the proceedings will be conducted.

Short hearing procedure (rule 7)

Under this procedure, there is limited time available to the parties to orally address the arbitrator on the matters in dispute. The time can be extended if both parties agree. If there is no agreement, only a day will be allowed for both parties to be heard. Before the hearing takes place, it is important that each party provides the arbitrator and the other party with a written statement of claim, defence and counter-claim as appropriate. Each statement must be accompanied by all relevant documents and any witness statements on which it is proposed to rely. The JCT procedures usefully insert some time-scales for certain of the steps.

This procedure is suited to disputes which can be decided fairly easily, perhaps by a site inspection, and of course the arbitrator has full powers to make such inspections as are deemed necessary. The arbitrator has a month to make an award after hearing the parties. Expert evidence is possible, but usually dispensed with under this procedure. Usually, the arbitrator's own expertise is sufficient and the parties can agree that the arbitrator use that expertise. Rule 7.5 prevents any party, which chooses to call expert evidence, from recovering the costs of doing so unless the arbitrator decides that the evidence was necessary for the decision – a strong disincentive. This procedure is suited to many simple disputes, which require some brief opportunity for a hearing and/or an inspection of the site.

Documents only procedure (rule 8)

This can be extremely useful, quick and inexpensive under the right circumstances. All the evidence required for the arbitrator to come to a decision must be contained in written form. Often this procedure cannot be used because of the need to present some oral evidence, but if the sums at stake are modest, there is much to recommend it. Each party must provide the arbitrator and the other party with a written statement of case giving details of the relevant facts and opinions and, importantly, the redress sought. The evidence of witnesses as to fact can be provided in witness statements, which must be signed by the witnesses concerned. Expert evidence may be given in the same way. As usual there is a right of reply to the claim and to any counter-claim.

Despite the title of this procedure, the arbitrator sometimes opts to question the parties and/or their witnesses at a brief hearing, but without examination or cross-examination. The arbitrator must make a decision within a month or so of the conclusion of the exchanges, but there is provision for the arbitrator to notify the parties that more time for the decision will be required. The JCT procedures again set out a useful timetable.

Full procedure (rule 9)

If neither of the other two options is thought satisfactory, CIMAR makes provision for the parties to present their cases in a manner which echoes conventional court proceedings. This is extremely useful for complex disputes where oral evidence is crucial to the arbitrator's understanding of the parties' positions and to the eventual award.

This is the most complex process and the JCT procedure which sets out a detailed timetable for various activities within the procedure is of real assistance to the parties and to the arbitrator. It is intended that the rules will accommodate the whole range of disputes which might arise, offering the opportunity to hear and cross-examine factual and expert witnesses. Therefore, they represent a sensible framework in which to carry out the process. They may be changed as desired to be used effectively and efficiently for the particular dispute under consideration.

The unamended rules lay down that parties will exchange formal statements. In difficult or complex cases, the statements will be similarly complex, incorporating many facets, for example, the claim, defence and counter-claim (if any), reply to defence, defence to counter-claim and reply to defence to counter-claim. The statement must set out the facts and matters of opinion which are to be established by evidence. It may include statements concerning any relevant points of law, evidence, or reference to the evidence to be presented. The arbitrator should give detailed directions concerning everything necessary for the conduct of the arbitration, often including directions regarding the time for requesting further and better particulars of the other party's case and the time for replying. The arbitrator may direct the disclosure of any documents or any other relevant material which is or has been in each party's possession. The parties may be required to exchange written statements setting out any evidence that may be relied upon from witnesses of fact in advance of the hearing. There will be directions regarding expert witnesses, the length of the hearing or hearings and the time available for each party to present its case.

Under CIMAR rule 2.1 and section 14(4) of the Arbitration Act 1996, arbitral proceedings are begun in respect of a dispute when one party serves on the other a written notice of arbitration identifying the dispute and requiring agreement to the appointment of an arbitrator. That is important for two particular reasons. The first is that it is relevant in terms of the Limitation Act 1980. If the notice is served before the expiry of the period, it prevents the respondent using the limitation defence. The second reason is that notice of arbitration may provoke a counter-claim from the respondent. Usually this is absorbed into the existing arbitration, but strictly, it is doubtful whether a counter-claim can be brought within the jurisdiction of the arbitration, which has already begun, without formal process.

A claimant wishing to delay or even thwart the respondent's attempts to bring a counter-claim into the proceedings may be able to do so. CIMAR rule 3.2 allows any party to an arbitration to give notice in respect of any other dispute and if it is done before the arbitrator is appointed, the disputes are to be consolidated. Rule 3.3 allows either party to serve notice of any other dispute after the arbitrator has been appointed, but consolidation is not then automatic. Rule 3.6 of CIMAR makes clear that arbitral proceedings in respect of any other dispute are begun when the notice of arbitration for the other dispute is served. There are serious practical issues to be considered.

A doubt concerning whether a counter-claim has properly been brought within the original arbitration may affect the extent to which either party has protection from liability for costs. This is especially the case if previous 'without prejudice' offers of settlement have been made. If the respondent wishes to make a counter-claim some time after the original arbitration notice has been issued, the costs may be high and the limitation period may have expired.

An advantage of litigation is that claimants can take action against several defendants at the same time and any defendant can seek to join in another party who may have liability. This facility is not readily available in arbitration which is a private arrangement, usually confined to the parties to a contract.

Rules 2.6 and 2.7 provide that where there are two or more related sets of proceedings on the same topic under different arbitration agreements, anyone appointing an arbitrator must consider whether the same arbitrator should be appointed for both. In the absence of relevant grounds to do otherwise, the same arbitrator is to be appointed. If different appointors are involved, it is required that they consult one another. Sensibly, if one arbitrator is already appointed, that arbitrator must be considered for appointment to the other arbitrations.

The situation usually occurs when there is an arbitration under the main contract and also between the contractor and a subcontractor about the same dispute. It is also possible that there are two contracts between the same two parties and an issue arises in both which is essentially the same point. It is good practice and usually less expensive if the same arbitrator is appointed for that situation.

Figure 14.4 shows an outline of adjudication and arbitration in simple flowchart form.

14.4 Legal proceedings

Article 8 deals with the legal proceedings option which provides that the English courts have jurisdiction over any dispute or difference arising out of or in connection with the contract. If the parties wish to use this, they must delete the arbitration option (Article 7). However, the default position has changed under this contract. If neither arbitration nor legal proceedings is deleted, legal proceedings are the default position.

14.5 Summary

- Adjudication can be used by either party at any time even if litigation or arbitration is in progress
- Adjudication is only temporarily binding
- It is a short and pretty rough-and-ready procedure

Figure 14.4
Flowchart of adjudication/arbitration

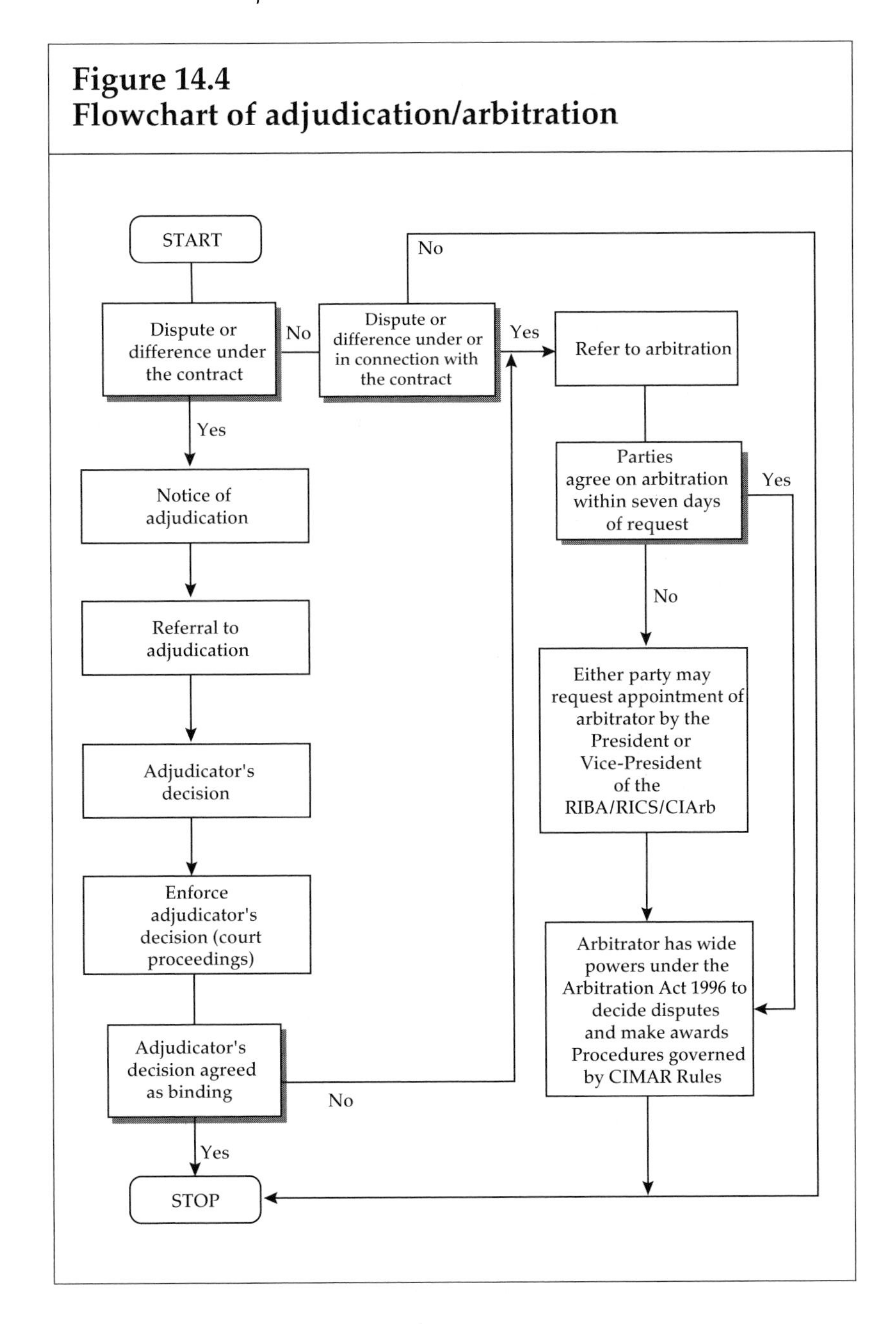

- There is no 'appeal' against an adjudicator's decision. If referred to arbitration or the courts, the dispute will be heard again from the beginning
- The courts will not overturn an adjudicator's decision just because it is wrong
- Arbitration can be commenced by either party at any time
- The parties should attempt to agree an arbitrator
- Arbitration is to be conducted under the JCT/CIMAR Rules and the Arbitration Act 1996
- If the parties have agreed arbitration as the means of binding dispute resolution, a party can be prevented from seeking litigation by reference to section 9 of the Act
- The courts' powers equal those of an arbitrator
- Legal proceedings are the default option if no procedure is entered by the parties in the contract particulars
- All forms of dispute resolution should be the last resort. Significant costs are incurred by both parties, even by the successful party
- There is no such thing as a cast-iron case and the words 'we can't lose' usually presage failure.

Table of Cases

Clause Number Index to Text

Subject Index